Anonymous

**Uses of Mental Science**

Anonymous

**Uses of Mental Science**

ISBN/EAN: 9783337816964

Printed in Europe, USA, Canada, Australia, Japan

Cover: Foto ©berggeist007 / pixelio.de

More available books at **www.hansebooks.com**

NUMBER     SEVENTEEN.

# THE
# HUMAN-NATURE
## + LIBRARY +

### DEVOTED TO WHAT MOST CONCERNS!
## BODY AND MIND

EDITED BY

**NELSON SIZER.**
President of the
AMERICAN INSTITUTE
OF PHRENOLOGY.
AND

**H. S. DRAYTON, A.M., M.D.**
Editor of the
PHRENOLOGICAL JOURNAL.

QUARTERLY.
10ᶜ a number.
30ᶜ a year.

## USES OF
# MENTAL SCIENCE.

WITH

## ADDRESSES

DELIVERED BEFORE THE

## AMERICAN INSTITUTE

OF

## PHRENOLOGY,

1890.

**FOWLER & WELLS Cº. Publishers,**
**775 BROADWAY, NEW YORK.**

# THE
# HUMAN-NATURE LIBRARY.

No. 17.       NEW YORK.       JANUARY, 1891.

## USES OF MENTAL SCIENCE.

A little more than a hundred years ago, that eminent physician of Germany, Dr. Francis Joseph Gall, promulgated a new method of studying mind.  Instead of treating it as it had previously been done, in the abstract and unconnected with physiology, he presented the subject in a new and startling light, maintaining that different faculties of the mind and character are manifested through special portions of the brain.  The subject of mental philosophy previously had been mainly in the hands of theologians, studying mind as they did from a theological point of view.  Dr. Gall made a new departure, and so eminent a man was he in his medical profession that he was appointed physician to the Emperor of Austria, and while thus related to his sovereign, he commenced his teachings and demonstration respecting the physiology of the brain, which attracted wide attention and awakened intense interest.

Dr. Gall and his chief associate and pupil, Dr. John Gaspar Spurzheim, travelled and lectured extensively in Germany, and in 1805 they visited and settled in Paris, and were everywhere received with marked attention and interest, and the best scholars and most influential men of those countries became ardent advocates of the new mental philosophy.  So profound an impression have the discoveries and teachings of Gall and Spurzheim made on the public thought, it may be stated on the best authority that in Germany to-day the relation of physiology to the brain and mind is everywhere appreciated, and the scholarly thought is largely tinctured, if not saturated, with the ideas which Dr. Gall taught.

ENTERED AT THE POST-OFFICE AT NEW YORK AS SECOND-CLASS MATTER.

I

We know that the writers of all countries, in reference to the
mind, talent, propensity, passion, virtue, criminality, and insanity, have enlarged their field of vision and modified their mode of
treating these subjects, covertly it may be in many instances, yet
manifestly the teachings of Gall run through their writings.
Teachers, at least the more advanced ones, are studying mind and
trying to instruct and guide youth according to the principles of
phrenology. Novelists describe their characters as having "a
low and villanous head," or "a broad, intellectual brow," or "a
head indicating dignity and firmness." They not only find the
phrenological terminology more clear and pertinent, but the easi-
est and best method of impressing their thoughts upon the
public mind. Some of them, of course, have tried to cover it up,
but those who are familiar with the subject can trace it as clearly
as the dog can trace the pathway of the fox where he has, ap-
parently to men, left no impression. It is amusing sometimes to
see how anxiously persons have sought terms that would not
show the source of their thought or of their information. Other
writers squarely talk the truth as revealed by phrenology, and
their writings are crisp and impressive in proportion as they are
soundly phrenological and pointedly and bravely stated.

Legislation is also influenced by the light which phrenology
throws upon human life and human character. Criminal jurisprudence has been modified, as well as the treatment of the insane, and everywhere there are gratifying evidences that mind is
being better understood under a higher and better law of exposition.

Phrenology, or the true mental philosophy, is doing for metaphysics, legislation, education, and jurisprudence, what the dis-
covery of microbes has done for the healing art. Each has
opened a better understanding of the nature of the subject under
treatment, and outlined a course of administration adapted to
the necessities of the case. Before phrenology was discovered,
people treated mind, as it were, in the dark. They inclined to
treat all alike, and wondered why the results were not equally
favorable. In medicine, before the knowledge of microbes, disease was a mystery and its treatment blindly empirical.

In theology, especially in sermonizing, fifty years ago a man
felt that he had struck the right chord when he spoke of the
corrupt heart of human depravity, with no effort at specification.
Now ministers who have read phrenology treat human tempta-
tion and human weakness and criminality in a way that people

better understand. If the prophet Nathan had preached a sermon to King David on abstract depravity, if he had discoursed eloquently on the exceeding sinfulness of sin, the king would have stood sturdily and agreed with him in every sense; but when he began to specify, so that the listener appreciated the point at issue, and the prophet said, "Thou art the man," the preaching began to have effect.

The Rev. Mr. Benton, of Vermont, a Congregationalist, stated to the writer, in 1842, that when he was in college and previously he was an infidel, or was so deeply tinctured with it as to conscientiously promulgate his views at every opportunity. Having heard phrenology called a species of infidelity, he procured several of the standard works, including Combe on the "Constitution of Man," for the purpose, if possible, of gleaning from phrenology something to bolster up and enable him successfully to defend his infidel notions. "But lo and behold," said he, "before I had half completed the first volume, my infidelity had vanished before the profound elucidations of the true mental philosophy. The same result occurred with a fellow-student in the simultaneous perusal of another volume, and we came out converted from our infidelity and are now ministers of religion, and I thank God for the timely opportunity of perusing those phrenological works. And I rejoice in an opportunity of opening to you my church and pulpit in which to preach phrenology."

The next Sunday morning, the course of lectures being only half completed, the minister took for his text, "Let us lay aside every weight, and the sin which doth so easily beset us, and let us run with patience the race that is set before us," Heb. xii. 1. He took up combativeness, destructiveness, acquisitiveness, secretiveness, and approbativeness by name and discussed their use and abnormal activity, and the people were marvellously interested. At that date ministers were less likely to talk phrenology by name than they are at present.

Phrenology has made progress: it has done more good than the world knows, remodelling, as it has done, literature and education, and modifying domestic and public administration of government.

It is believed that the addresses and papers read at the closing exercises of the American Institute of Phrenology, October 11th, 1890, showing the opinions of the students and the benefits derived by them from the study of human nature, will be of wide and abiding interest.

The class, which consisted of forty-two members, including ten ladies, one minister of the gospel, two doctors, one lawyer, and many teachers, was marked by intelligence, good character, earnest love for phrenology and its kindred sciences, and decided social harmony and good-fellowship.

The president, Mr. Nelson Sizer, in opening the exercises, said: On all former occasions of this kind the teachers have spoken first, followed by the graduates. To-day it has been decided to ask the students to come forward, and to give them a clear field. Afterward the faculty may be heard.

I am informed that the students have selected several of their number to speak in behalf of the class, the first of the list being Mr. George MacDonald, who, though a good American citizen, is a worthy son of that brainy land which gave to the world so eminent a mental philosopher as George Combe, and Brother Mac-Donald earnestly adopts his teachings.

## ADDRESS OF GEORGE MACDONALD.

Mr. President, members of the faculty, and fellow-students: Last night, at the banquet of the alumni, our venerable chairman said that the exercises which were then being held were to be recorded in history, it being the celebration of the first quarter century of the American Institute of Phrenology.

In this class the Alumni Association has been formed and its foundation laid.

It is therefore well that on this our first foot-hill of our historic Pisgah we should view the landscape, even if it be in a cursory manner.

When the immortal Gall first ascended the public platform and gave to the world the statement of his observations and discoveries, there were opinions, the most derogatory and erroneous, promulgated by professional reviewers and critics concerning the new science; producing the belief in the minds of many that it was the effusion of a bewildered imagination which would only be accepted by ignorant and weak-minded people. For many years, in fact up to recent times, such pernicious sentiments had to be met and "downed" even among high and classic scholars. But, thanks to the better intelligence of the people and their thirst for knowledge and truth, the ignorance, errors, and prejudices of those days have melted away like the mists of the morning in the rays of the sun of science and reason.

The battle-days of phrenology are over; those days when Spurzheim and Combe, and Caldwell and the Fowler Bros. and Wells, and many others had to grapple with the master-minds of the world. Theirs were the days of battle and clearing; ours are of peace, of sowing and reaping.

Those pioneers were observers, thinkers, and philosophers

they were mostly men of strong convictions, and as such, of course, had their followings; and although phrenology is a well-defined science, still there were differences of opinion as to certain manifestations, and definitions, and temperaments, and names of organs, etc.

In Europe and America societies and classes were formed from time to time for the promulgation of the science, and many bright, practical phrenologists sprang from them, as well as many ignorant pretenders. But up to the time of the founding of the American Institute of Phrenology there was no place that the world could call The Universal Home of Phrenology; where it could be studied as a science along with all the kindred branches relating to the study of mind as connected with the body. There was no place where the precious relics and archives of the science could be stored and preserved.

This institution has nobly filled that mission. It has been patronized to a remarkable degree. Great Britain, Continental Europe, Australia, New Zealand, South America, Canada, and every State and Territory in the Union have sent it students. The large and unique cabinet of skulls, casts, drawings, and scientific apparatus, combined with its scholarly and efficient staff of professors, afford unequal facilities to students of brain and mind.

Look at the collection of skulls and casts. Every one of them has its own peculiar history; each is its own record of the deeds it has done and *how* it did them. Those skulls and casts of noted characters, if destroyed could never be replaced. The souls who shaped them have gone to God, and left with us the testimony of what was and what might have been. If some one would write up the history of the incidents and circumstances connected with the procuring of many of those specimens, it would form a brilliant chapter in the romance of phrenology. There is also a large collection of portraits and drawings; some of them are unique, others are rare and valuable; all having a scientific significance which the student of mental science can at once appreciate.

The literature given to the world under the immediate stimulus of this institute is large and valuable. "Heads and Faces" has already reached its eighty-fifth thousand, and in a few years it is destined to reach a quarter of a million copies. "How to Read Character," "Choice of Pursuits," "New Physiognomy," "Popular Physiology," "Brain and Mind," and scores of other works. Those I have named have emanated directly from the professors and students of the institute. I may be pardoned if I make special mention of "Brain and Mind," as I think it the brightest jewel in the diadem of this institution, being the work of two of her children. That book for clearness of definition, accuracy, and illustration is matchless. It will stand as a monument to honor the names of Drayton and McNeill for years to come.

Before reviewing the living functionaries let us first do homage to those who have passed beyond—Samuel R. Wells, whose labors and sacrifices, along with some who are present, laid the foundation of this institution. We should also remember the name and labors of O. S. Fowler, whose writings and industry in

the field of phrenology have done much to make it a familiar
science among the American people.

We shall now take a glance at those whose work it is to guide
this craft through calm and storm, and I think we will all say
that everything is in perfect harmony and each well fitted for
their respective offices.

We shall first mention Mr. Albert Turner. We have not yet
conferred a "professorship" on him, but when we do I think it
should be "Professor of Our Material Wants." I am sure that
each of us individually, and the class collectively, feel deeply indebted to this kind gentleman for his great foresight in making
arrangements for the comfort and economy of the students. His
great heart and mind have fitted him to a wonderful degree for
his responsible position in connection with this institute and the
Fowler & Wells Company.

Dr. N. B. Sizer, who fills the chair of anatomy, physiology,
and diseases of the body and brain, is a finished master-scholar.
As a lecturer and demonstrator he has few equals. His analysis
and dissections, his deep microscopic research, his power to illustrate all subjects which come under his hands—in short, everything is treated in such a clear, scholarly, and common-sense
manner that his lectures are listened to with delight and fascina-
tion.

Dr. John Ordronaux, by his lectures on insanity and jurisprudence, has imparted to us a line of thought and means of observation which will enable us to pursue the further study of those
complicated and critical subjects with pleasure and profit.

We have been charmed by the lectures of Dr. Gunn on human
magnetism and psychology. He has demonstrated those myste-
rious forces in such a clear and reasonable manner that we can
now view it as a well-defined science.

The Rev. A. Cushing Dill has lectured to us on elocution and
vocal culture in relation to public speaking. They have been of
great benefit and will tend to give ease and grace combined with
self-reliance, accomplishments so necessary to those who are fitting themselves to talk before an audience.

The course of lectures delivered by Dr. H. S. Drayton have
been masterly and scholarly efforts. He has followed the lines of
thought pertaining to mental science from the remote Greek
philosophers, step by step, to those of the present time. His lectures on physiology and mental phenomena have been highly
edifying and instructive, imparting to those who have had the
good fortune to listen to him a mental discipline which will
leave its impress for life. It is fitting that we now take cogni-
zance and appreciate the great ability and effort he has exercised
in editing the *Phrenological Journal*, in forwarding and maintaining that high standard of excellence which it has obtained,
and it is our duty as the alumni of this institute to do all in our
power to forward the interests and prosperity of this our organ.

Once more we cluster around our mother in Israel, Mrs. Charlotte Fowler Wells. In her we have found the fostering mother's
care with kind advice and counsel. We have heard from her the
story of her girlhood and the eventful history of her after-life;

the story of the great pioneers in phrenology and her personal experience with them; of their failures and successes; of their personal appearance and peculiarities of body and mind. These things we have listened to with the raptured charm of children when hearing the tales of childhood at their mother's knee—tales and circumstances never to be forgotten. As Miss Fowler in her father's house and as a school teacher, she was early imbued with the grand principles of the new philosophy of Gall and Spurzheim, which her illustrious brothers had just espoused. While she was teaching school she formed a class, the first phrenological class on the American continent. When her brothers O. S. and L. N. adopted phrenology as a profession they chose her as their assistant in business, and a great acquisition she was to them, for when the days were darkest and the young men became disheartened, her hope and courage, her advice and foresight kept them at work till the clouds had cleared away. Thus she came in contact with the great champions of the cause in America, and also with some of its strongest opponents. The office of Fowler Brothers was the centre of phrenological gravity on this side of the Atlantic. While she was thus engaged that brilliant and energetic gentleman, Samuel R. Wells, became converted to phrenology and joined her brothers in their great work. He was a man of remarkable tact and winning manner, and he showed his sagacity in choosing Miss Fowler for his life companion, and they worked hand in hand through the ups and downs of fortune; writing and publishing, speaking in public and private, breaking the ground and sowing the seed. With such opportunities and vast experience, with such blending of temperaments and balance of brain, it is not extravagant to say that Mrs. Charlotte Fowler Wells was a great acquisition to phrenology and its promulgation in America. We trust that her autumnal sun may be long in setting, that its soft and mellow rays may long continue to shed abroad its rich and ripening lustre on the classes of this institution.

I think it is here meet to say a word of congratulation and thanks to our absent friends in England, for whom we cherish the greatest love and admiration for their industry and self-sacrifice in sustaining the work on the other side. We mean Mr. L. N. Fowler and his daughters. They have, through their admirable writings and lectures and their spirited magazine, carried phrenology into the baronial halls as well as the cottage homes of England. They have made a following of thoughtful men throughout the British realm, and it is with much pleasure and pride that we send them our greetings.

And now we bow to "the greatest Roman of them all," Nelson Sizer! Endowed with a solid, well-knit frame and a finely-balanced brain to govern it, God planted him in his favorite nursery, "the rock-bound hills of New England." Little did he think in his boyhood years, when climbing those rocky crags, that he was toughening and fitting his body for the fatigue and siege of a centenary. New England was poor in those days; its boys had to work, and work hard, for existence; the soil being hard and stony there was little return for the labor; every dollar was saved by close and rigid economy. Such life strengthened their bodies and sharpened their wits; their reflectives and perceptives grew; everything developed that was needed in the economy of life. These are the schools and schoolmasters that have made New England great and made her power to be felt among the states and nations of the earth. It was there that Nelson Sizer began to obtain that masterly insight to human nature and human thought that is now wielding a power and regulating the lives of millions of the human race through his tongue and pen. His university course was taken mostly by the big New England fire-place in the intervals of labor. His professors were Gall, Spurz-heim, and Combe; and he knew every word they had written on the brain and mind. When he entered the field of phrenology as a profession he had no classic lore with which to embellish his subjects, but he was well primed with the spirit and teachings of his masters. I question if ever there was one started out in the lecture field better grounded in the teachings of these three mas-ters than he was at that time. The ten years occupied by him in lecturing and examining gave him a world of experience in fitting him for his after-life in New York. All his life, and to-day more than ever, he is a monument of industry. Look at the work he has done and is doing; his authorship; his articles for the *Journal* and other papers; his work in phrenological exam-inations—in this capacity he has had two hundred and fifty thousand heads under his hands during his professional life. Look at his work as professor in this institute: he delivers two and sometimes three lectures a day during its entire course, be-sides filling up the intervals with professional examinations in the office; and he often commands time to talk with and advise a student outside of class hours. We are sometimes asked, "What is the secret of his power?" It is no secret; he preaches it from morning till night every day of his life; it is his life's work, telling people how to live and how to act toward them-selves and their fellows; he tells them how to use common sense and the instincts which God has implanted in all his creatures; he tells them their weak points and how to improve them; their strong points and how to make the best of them. He turns a man inside out, as it were, and gives him a view of himself, and shows him how to live to make the best of life for himself and others. As an examiner he stands supreme. Gall, Spurzheim, Combe, Caldwell, Fowler, and Wells were all masters of phrenol-ogy each according to his talent; they were all independent thinkers, observers, and discoverers; and Sizer has culled and stowed away in his fertile brain the best of what they knew, and adds to that the results of his own large experience and innate genius. His invention of the "New Facial Angle" is of infinite value; his work on "Resemblance to Parents" is essential to the phrenological student, and his doctrine of "The Proclivity for Co-operation of the Organs," besides being a quick and decisive method of arriving at facts, has imparted to phrenology a wider range of vision than it has heretofore possessed. His application of physiology, the laws of being and of health, in his examina-tions, has elevated phrenology as a science and widened its scope of usefulness under his hands. The moral glory and strength of Mr. Sizer is in his consistency; he acts what he preaches. For more than fifty years he has talked and written to men concerning their health, and being, and usefulness, and it is safe to say that at least five millions of people have read something from his pen; and if his teachings have been accepted and acted upon, the longevity of the race will be much increased. I have said that he practises what he preaches. Just look at him: at nearly seventy-nine years shows no loss of teeth, is as active and strong physically as the majority of men are at forty, and mentally he can put to blush the brightest student of any of our classes. Barring accident, I have little doubt but what he will see his hundredth year. We can point to him and also to Mrs. Wells as living testimony to the truth of phrenology and the work it will accomplish for mankind. Considering the vast work for good which he has accomplished during the past half-century, the popularity with which his writings are being received, his vigor of body and mind at this day, his name and fame should pass into history in the first ranks of the benefactors of the world.

Now, my friends, I have given you an inventory, although it is a very imperfect one, of the stock-in-trade, of the American Institute of Phrenology at this date; and I think it will stand comparison with any scientific school of its age in the world to-day, both as to material and brains and the number and quality of the students it has graduated.

The Alumni Association of this institute may well be proud of such a nursery; we have "sat at the feet of Gamaliel." Let us therefore tell the world what we are and where we have come from. Let us now bend our energies to gather material to build a home, a permanent home, a universal home for phrenology. Let our mystic seven (the Groups) join together in holy harmony and work with diligence till that end is accomplished. Let it be a monument in future ages to those who lived and worked for the benefit of the human race.

---

## ADDRESS BY MISS VIOLA EVANS.

Mr. President, ladies and gentlemen: The word phrenology has sounded in my ears since the days of my babyhood up to the present time. All I knew of it then was that it was a long, hard word for me to pronounce, yet I think I loved it then. I know I do now.

The firm name "Fowler & Wells" was familiar to me as my own, being among the first I learned to speak when a child. God grant that the name may continue to live and prove as great a blessing to future generations as to the past.

I first gained my love for phrenology through a brother who became interested in the science when a boy, procured some books on the subject which were read with much interest in my father's family.

Since hearing of the American Institute of Phrenology I have had an intense desire to become a member of that body of men and women and to receive such instruction as would enable me

to go out into the world and teach to humanity the ways of life
and happiness and thereby render service to "Him who went
about doing good."

Our dearly loved science, phrenology, shall live.  Let us go out
into the field with the full determination of being successful.  It
is right that we should be successful.  But what is the standard
of measurement?  There is none save the one given us eighteen
hundred years ago, a success too great to be measured, and yet
the unit of measure of all human lives.  The world has false
scales by which it often measures success.

But let us be men and women in the work.  It is a noble thing
to be a man!  It is a noble thing to be a woman!  Then let us
make every exertion to be something and do something in this
life; let it be our dearest hope and fondest prayer that we shall
be noble, pure-minded men and women wherein doth dwell the
spirit of God.

For six weeks we have been drinking at this fountain of truth,
yet we are not satisfied; we are glad of this thirst after more
knowledge of one of the grandest sciences that has been brought
to mankind; and he who does not thus feel ought to be ashamed
to bear the name phrenologist.

Let us go out in the world with the stamp of truth upon us,
feeling that we as phrenologists must set our mark high, work
with clean hands, for we are dealing with one of the noblest
works of God.  I rejoice to know that we have so many brave,
energetic, devoted men and women who have taken up the science
of phrenology as a life-work.

It has been said by some perhaps, Phrenology is not true.  It
has also been said perhaps, Harvey's circulation of the blood is
not true.  Let us deal carefully with such, for ignorance is abroad
in the land.

It is the duty of the phrenologist, after having gained such
valuable information from this most honored faculty, to spread
the good seed wherever we go, ever bearing in mind that by one
misdeed he is dishonoring these venerable pioneers who have
stood the storms and fought the battles of fifty long years.

We want to be worthy the name phrenologist; we want to go
out from this place worthy to be called the children of the American Institute of Phrenology.

It is true we will meet with many discouraging words, many
a critic's eye will be fixed upon us, for the phrenologist is supposed to be perfect.

Although the way may seem dark now, dark on through the
future, we must remember that light shines from the heavenly
turrets.  Disappointments will come to us, but I presume life
brings to all some Dead Sea apples.

Brethren, this is a day of rejoicing, yet it is a day of sorrow.
We rejoice because we have been permitted to sit and drink in
these wonderful truths and become filled, as it were, with information that could be gleaned from no other field.  We are
filled with sorrow at the thought of parting; we have been associated as brother and sister and have learned to love each other
as such.  But now we must say farewell.

## ADDRESS BY RICHARD SMITH.

Mr. President, members of the faculty, ladies and gentlemen of the class of 1890: The occasion which calls us together now, to say a few words of congratulation and encouragement to each other ere we part, naturally makes us feel that life is fleeting.

Human life, conscious existence, with its modes of expression and the why therefore; the why this or that action; why differ we, has drawn us from all parts of this country and Canada to hear what phrenology has to say with its mass of facts bearing on human action. Our teachers have all borne patiently with us and our many queries, which to them must have seemed like lack of attention on our part, for we all agree that the wheat has been given us without stint. Faithful souls with ready speech and willing hands and ears have opened the rich storehouse to us, not satisfied to teach us only that man has a head, but asks us to go inside and see the wonderful mechanism and follow on down and see that the "Man Wonderful" cannot say to any member, "We have no need of thee."

In Professor Nelson Sizer we recognize the wisdom and facts of phrenology collected on many well-fought battle-fields, the spoils from which, loaded down, he readily and willingly imparts to all who come to him; grown old in the service yet yearning for more light; and happy in the service, we wish him many more years that he may see the glorious truths take greater hold on the public mind.

Mrs. Charlotte Fowler Wells, from whose lips we have heard the history and the trials of phrenology, who with Professor Sizer forms the strong hands which, reaching up to the source, are passing down the fruit to us children for the future. May God bless them, and may the object so dear to the heart of Mrs. Wells be fairly granted before she be called home.

Dr. Drayton's able exposition of brain function, its offices, have certainly been a revelation to us all, and may we show our appreciation by profiting by his teaching.

Dr. Sizer deserves our thanks for the amount, kind, and manner of anatomy and physiology he set before us, also for the beautiful brain and the patient manner in which he dissected and explained the same to us.

Dr. Gunn we will remember gratefully for his pleasant explanation of hypnotism and psychology.

Dr. Ordroneaux has left us indebted to him for his masterly and clear treatment of insanity and jurisprudence. We thank him.

And last, but not least, we will remember with gratitude the Rev. A. C. Dill for his wholesome advice and his earnest endeavor to teach us how to use the facts and knowledge here gained so as to present them to the public in an acceptable manner.

And now, classmates, let us take these truths to our various fields of labor, and whatever may be our vocation, let us not forget the light we have received here has a claim on us. It will be no less, but increase if we lighten our neighbor.

Let us be hopeful and earnest, remembering the desires of the American Institute of Phrenology. Sowing the good seed everywhere. Do not be discouraged, my classmates: if the mantle of Elijah has not fallen on you, neither will the service of an Elisha be required. Let us recognize the truth of phrenology in ourselves as in others, and speak the truth. Learn to labor and to wait, going out in strong yet simple faith like the children who sing,

> We are little sunbeams
> Singing as we shine,
> You in your little corner,
> And I in mine.

## ADDRESS BY MISS PEARLE BATTEE.

"Our duty as teachers and responsibility as examples in the sight of God and mankind."

Mr. President, members of the faculty, and fellow-students: The time has come and now is when we must bid adieu to each other and start, as it were, afresh on the highway of life, each one, we trust, better able to cope with the struggle which must come to all than he was at the opening of this institute; for we have here had such instruction as we have never had before, and with knowledge comes *ability* and responsibility.

How many of us realize that as we go forth from this well-spring of truth and knowledge, whether at home or abroad we will be looked upon as instructors and examples, will be criticised and copied. Then if this be true, how great will be our responsibility for our every action in life.

We have heard it remarked that we must raise phrenology to a high standard and put it on a firm basis; this we need not waste our energy trying to do—it is already there; it is the highest, noblest, best and most practical science and means of accomplishing good known to man. Then what are we to do? This, brethren: raise ourselves to such a standard that we shall be *worthy* the profession we have chosen, and make ourselves such that the honorable faculty of the American Institute of Phrenology may *proudly* claim us as their children.

The material upon which we have chosen to work is God's work, his ablest, highest, most exalted, perfect, and divine; because it is not only the most perfect matter, mechanism, and intellect, but greatest and more than all else in his universe is the human soul divine. Who can stand in his presence and guide and direct this immortal soul and build up this temple, the body, which it inhabits, without a feeling of awe and responsibility and without a desire and determination to attain toward perfection, to study to show himself approved of God, a workman that needeth not to be ashamed? *When* we so live and conduct ourselves, and *only* then, will the outside world look upon phrenology in its true light; for knowing nothing of the science, they will judge it as its dissemblers, in their lives as well as their teachings, represent it to be. Therefore, brethren, let us go where duty calls us and there hold up our pure, bright phrenological light and make it so to shine that others may see our good works and honor, respect, and adopt this ennobling truth. And if we all live and conduct ourselves as if we felt each day of our lives, " Thou God seest me," then we cannot go far wrong.

For four long years I have had my mind made up to attend this course of lectures, and all that I have done in that time has been gravitating to this point. I say long years, for I have stood in alternate fear and hope of my ability to succeed, and had I been assured and made to feel that I would not realize my ambition, I should have been one of the most miserable, for from my first-accurate knowledge of the science of phrenology, I have said in my innermost heart, I will consecrate my life to it, for I have known nothing so good, great or worthy of my honest effort to glorify God and benefit mankind. Nothing less would ever have satisfied; for my whole soul approves and my heart is in it, and though I feel weak and trembling of myself, yet in His strength and the cause of phrenology I go forth strong.

I have been made happy and stronger by the many kind wishes expressed by the faculty and my classmates for my future success, and I cannot refrain from quoting some lines that were composed for me to use as a text in my life-work, and I desire to sincerely thank the composer, Mr. George MacDonald, for it; and I give it to you all as my best wish for your future success.

> Lift up humanity;
> Cleanse it from vanity;
> Waft all around it the zephyr of love;
> Show it where manhood is.
> There's truth and righteousness;
> Lead it with placid pace, onward to God.

## ADDRESS OF DR. JOHN C. BATESON.

Honored president, professors, and fellow-graduates: With joy and sadness commingled I stand before you. Sad because of the thought of parting so soon from this the American Institute of Phrenology, where homelike many a pleasant hour has been spent imbibing choice wisdom from an inexhaustible fountain. Sad because of the parting from kind and devoted instructors, who have taken every pains to make the session pleasant and profitable, and to inspire in us a true love for and knowledge of the course to which they as specialists are giving their lives. As followers we shall often miss them when we come to battle against waves and storms of discouragement that will confront us out in the great sea of humanity. We shall wish for their skill and experience to aid us in time of need. Again, I am sad because of the thought of separating from fellow-students to whom I have become attached through sympathy with them in a common work. They, too, will be missed. The many social interviews, the kind regards, and the friendly feelings expressed will long be cherished with tender recollections.

On the other hand, I am joyful for having had the privilege of completing a course of study in this great scientific school of human nature, the only school of the kind in the world where a thorough training in the intricacies of character-reading can fully be obtained. Again, I rejoice in the opportunity of meeting so many champions and friends of phrenology, especially our venerable leaders, President Sizer, Mrs. Wells, and Dr. Drayton. They are to us as father, mother, and elder brother. The advice and instruction received from them can never be calculated in true worth. And I am sure we have all enjoyed the masterly and instructive lectures delivered by Drs. Sizer, Gunn, Ordroneaux, and Prof. Dill. They have rendered us much needful information. And furthermore, it is a satisfaction to know that there is such a lively interest being taken in the science that is destined to teach mankind how to observe the great injunction, "Know thyself," and how to make the most and best use of mental and physical talents. The fact is that phrenology, rightly understood, is no illusory fancy or notion, but serves as a key with which to unlock the door of the human mind, and there is no other system of mental philosophy whereby we can gain a better acquaintance with the action of mind upon the body, and that will direct us how to perceive the subtle relations between mind and certain parts of the body. Many diseases arise from morbid conditions of the mind-organs, hence the necessity for mental treatment.

The physician with a knowledge of phrenology can know better how to adapt himself to the disposition of his patients in giving suitable advice and words of comfort. By recognizing the temperaments, he can thereby know whether to administer large or small doses of medicine. There are hundreds and thousands of persons with large brain powers going down to their graves unheard of and unsung, just for the want of phrenological guidance. Could they be placed in the proper channels and trained accordingly, the probabilities are that many intellectual lights would be shining where now all is dark and void. In the hidden waters of the great Mammoth Cave of Kentucky there are small fish with eyes, but they see not; the optic nerves are wasted away till there is no power to exercise the function of sight. This is the result of being kept in a state of darkness, where the optic apparatus is perpetually unused.

From this example we learn a lesson, viz., that whenever a living organism is prevented in any way from performing its proper functions it degenerates and dies. In the human brain there is undoubtedly a large number of mental faculties that are lying dormant or withering on account of not being educated for use.

Then as phrenologists and physiologists, we should instruct people how to exercise their mental gifts to the best advantage, and thus save them and their posterity from the penalties of neglected and violated law. The dissemination of phrenological doctrine by conscientious teachers is greatly needed in any civilized land—teachers with a missionary spirit who love the cause for the sake of accomplishing good rather than for the mere gain in money.

False and disreputable teachers have done untold harm to our noble science, but we must build up that which is broken down, and overcome every reproach by our good example and precept.

And when our earnest endeavors have proven tried and true, then we shall reap the fruits of our labors.

In order for us to accomplish success, we must not stand and shrink and think of the risk and danger, but go forward with brave hearts and strong determination to win victory.

And now I must say farewell. Let loyalty and honor ever rest upon the class of 1890.

## ADDRESS BY MRS. CLEMENTINE G. HENRY.

Mr. President, honored faculty, and classmates: Some wise-acre has defined the three essentials requisite to round out a human life to be the following: good health, a good conscience, and an object in life. I fail to recall what prescription he recommended for the attainment of these several treasures, but I think phrenology is entitled to say, " Eureka—I have found it," for it is the royal road to all three of these choice blessings. Our pet science, and I hope you all feel that it is intrinsically yours, goes to the root of things and teaches, among other matters, how to regulate the social relations with the object of starting children in life with healthy constitutions. If phrenology is called in only after some of its primary laws on this point have been violated, it, like a good physician whose mandates for preserving health have been neglected (although few of the professions are believed to devote themselves to this branch of the art), shows the way to obtain health by right living in general and diet in particular.

A good conscience depends a great deal, so to speak, on how much of a brake we are enabled to put upon our propensities and inclinations. Phrenology, like the fairy god-mother in the nursery story, was willing to help endow the child with a well-regulated moral balance, but it again depends a good deal whether the parents were so constituted themselves as to transmit an evenly-balanced mental organization. In case this was inherited by an individual, and he is in full possession of all the faculties with which God has endowed his noblest work, man,' not the least among which being free moral agency, it will depend whether he uses these talents to his weal or woe; whether the mental mirror of his life, conscience, is clear and transparent, or like a stagnant sheet of water, full of decay. If a person becomes entangled in the labyrinth into which his propensities have led him, and he by chance clutch the anchor of hope which phrenology holds out to him, he may yet be reclaimed and led back to the point where two paths diverge, and started in the right one.

The last of these treasures is an object in life. Happy those who have one. It is a lamentable fact that many are utterly void in this respect, and drift here and there like sea grass tossed from its home, the side of the wave-washed rock. Phrenology teaches how to concentrate mental and bodily endowments to attain the best aims. If business, the money-getting part alone, if any trade or profession be considered as an object in life, this, in the light of the higher aspirations of man, would be a poor bauble to devote forty or fifty years to, but I think that the spirit of phrenology on this point is that in selecting a profession—the law, for instance—it ought to be the object of its advocate to use his knowledge and talent only toward the moral and the good, and to make it his object to combat untruth and injustice wherever he meets it, and to champion only a deserving cause. A physician who has embraced his profession in the spirit of phrenology would use his power conscientiously and devote his life to the relief of suffering, verily a noble art, and not stoop to the tricks of the profession. If we consider a trade—printing, for instance—and if all in that business had made it their object in life not to print or circulate what may be injurious to a fellow-creature's welfare, a great step would be made in the advance of righteousness. If we look around us and make our eyes and ears bear testimony as to how the dignified title of lawyer has often become synonymous with the less euphonious term of rascal, when we hear of the shady source of revenue of many doctors who have studied the laws of nature to use them against that kind mother, and when we look upon book stands and see them filled with trashy literature and consider how talent, money, and time have been wasted, we then begin to realize what work there is left for phrenology to do. Much might also be said of trades and manufactures the products of which are injurious, such as the manufacture of cigars and tobacco, the distilling of liquors which involves the destruction and pollution of wholesome grain, for the want of which a whole nation may be starving in another part of the globe. The manufacture of certain so-called food products, which are no food at all, come under this same heading.

I hope as the class go out into the field they will at all times fearlessly urge it upon their subjects to choose occupations which are a benefit to their fellow-men and which are compatible with the teachings of true religion.

I think phrenology is a grand science, that we are to-day constituted missionaries to go forth and tell people how to apply grand truths to every-day affairs, which, after all, are the alpha and omega of our lives. I hope you will make it your thorough study to advise parents in regard to their children, always remembering that "the child is the father of the man," and that in advocating truth you sow a seed which will keep bearing fruit when you are no more.

I think this is some of the kernel in the nut of phrenology, and I wish you all a hearty appetite for this delectable feast of reason.

---

## ADDRESS OF W. E. HALL, M.D.

Mr. President, ladies and gentlemen: A few weeks ago we gathered at the American Institute of Phrenology from the East and West and from the North and South. We met as strangers, but we part as friends and co-workers. And before we take each other by the hand to say farewell, let us consider some of the duties and responsibilities that will devolve upon us as public teachers; and let us feel that our mission is second to none in all the earth.

In the past history of the world the men who have been the most successful in plunder, rapine, conquest, and murder in times of war were eulogized to the skies as if they were gods in human form. But as men become moral and intellectual, the less veneration they have for war or warriors. And let *us* hope that soon armies will disappear from the earth. The mission of the phrenologist should be to bring "peace on earth and good-will to men." And this peace on earth and good-will to men can only come to nations as well as individuals by studying each other's interests.

Our mission is higher than that of the physician, because the natural laws teach us that we have no business or right to be sick unless we gain that right through disobedience to the great laws of the Creator. It will be our mission, then, to teach obedience to the physical and organic laws, that men may be healthy and happy.

Our mission is higher than that of the lawyer, because the lawyer as well as the physician gets his gain through the misfortunes of others. When the phrenologist shall have taught the world the supremacy of the moral sentiments and intellect over combativeness and destructiveness, we will then have need of but few lawyers.

Our mission is still higher even than that of the minister of the gospel, because we are to teach not only morality and virtue, but that the physical and organic laws are divine and to transgress them is as much a sin as to transgress the moral laws. A condition of civilization that causes one-half of all the children to die before the age of five years and a majority of the other half before they have lived out one-half their three-score years and ten, teaches us the sad lesson that there is more ignorant than wilful violation of the great Creator's laws.

Only the wise phrenologist can or will tell to the world the cause and apply the remedy, which is to roll back this great array of sorrow, sickness, transgression, and death.

· These causes may be found in men and women marrying who are mentally and physically unfitted for each other, and seem to live only to make themselves and others miserable. Again, others marry who transmit to their children criminal tendencies and the very germs of disease and death. Other thousands die because of unsanitary homes, poor food and raiment.

In the future as in the past, to whom, if not to phrenologists, are we to look for self-help and self-culture and self-knowledge? Wherever we find people believing in and practising the principles of physiology and phrenology, we find greater content with life and its duties, as well as living to a good and cheerful old age. It is with no little pride that we can point to all the great phrenologists, both the dead and the living.

Then upon us as teachers devolve the greatest responsibilities. We should always be careful students of anatomy, physiology, and phrenology. Yea, let us go to earth and sea and sky for knowledge, for we shall have need of all and more than we can

gain in this life. Then as we go out into the world of thought
and action we shall make men know and respect us. Let us
make others feel that the advice we can give them is a question
of usefulness and happiness to them in their business as well as in
their homes.

Never in all the history of the world has there been such an
eager, earnest, and feverish desire on the part of people to gain
wealth and knowledge as in this age of ours. Everybody we
meet is an exclamation or an interrogation point. How and why?
are the questions. What can I best do? How shall I marry?
How shall I train my children? How can I make money? How
can I make my home happy? How can I have good health?
And not infrequently the question will be asked, How can I
shun the duties of wifehood and motherhood? To all these
questions let us be prepared to advise intelligently. To the last
question we can only tell them that to be strong, willing, and
happy mothers is filling the woman's highest and holiest mission
on earth and filling the world with better men and better women.
If we intelligently teach what has been hinted at in the foregoing
lines we shall be, as it is our duty to be, moral philosophers as
well as phrenologists.

And of all that bright constellation of stars that has risen and
set forever, George Combe was the brightest of them all, and let
his moral philosophy and constitution of man be lights shining
upon our pathway wherever we go.

Great moral problems are now confronting us as individuals,
as a nation, namely: labor and capital and the race question. A
strict observance of the golden rule is the only solution of these
problems that we have.

In the past the Christian world has accepted the doctrines of
the Great Teacher in theory only, but have rejected them in
practice; when it is accepted in practice these troublous national
questions will drift away from our horizon like mist before the
morning sun.

Now, dear teachers, we feel that we cannot express gratitude
enough to you, but we believe in the coming years we shall appreciate you more and more. We shall often see your patient,
forbearing, and kindly faces; shall see your hands outstretched to
help us in times of trouble; shall hear your voices and feel your
presence wherever we go. And it is our sincere hope that when
you grow tired and lie down to rest there will be others to fill
your places with credit and honor; that the science you have
loved and taught so long may grow and spread all over the land.

> " We've been long together
> Through pleasant and through cloudy weather;
> 'Tis hard to part when friends are dear;
> Perhaps 'twill cost a sigh, a tear:
> Then steal away, give little warning,
> Say not good-night, but in some brighter clime
> Bid me good-morning."

The President: The class has so thoroughly presented the
merits of phrenology, and given the teachers and each other so
much pleasure in listening, that there seems little opportunity to

add anything. Nevertheless I will ask Mrs. Wells, the vice-president, to contribute something to the interest of the occasion.

## ADDRESS BY MRS. WELLS.

Children of the Institute of Phrenology of 1890: You are all our children—we have adopted you; we have other children scattered over the world. We have become attached to you. Since we have been together it is hard to say farewell, but we say God speed you on your way—may success attend you. I will not say, "Do not disgrace our institute," but I say, "Be a blessing not only to this institute, but to yourselves." If to yourselves you will be to your teachers. I am thankful that we have met; I am thankful that your responsibilities are increasing by knowledge. Your lives are sacred, more so now than ever before. Remember that the eyes of the world are upon you, as well as all who know of you, and do your best. It is expected you will take good care of your health on your way home from here, so that your families may recognize the improvement made since you came to the institute lectures.

We do not profess to keep a sanitarium, but your eyes are brighter, your color fresher, and your weight so much increased that your friends will think you have been sojourning at a health institution. Your minds have been fully occupied on healthful and interesting subjects; that even malaria, if you had it lurking about you when you came, has been eliminated by your healthful surroundings and congenial occupation; and you have learned and put in practice such lessons as, if faithfully followed, will keep you well and add length of days to your life.

Your teachers have enjoyed this season with the students, and will expect frequent reports from you of your success in doing good to others.

I will not longer claim your attention, except to express my good wishes that you may be a benefit to your race and ever recall with pleasure these happy days while attending the lectures of the American Institute of Phrenology for the year 1890.

## REMARKS BY PROFESSOR DRAYTON.

I feel, Mr. President and fellow-alumni, in the state of what might be called brain vacuity. This idea of permitting the students to speak first which has been inaugurated to-day has operated very unfavorably upon one of us. I think I must have recourse to a process that it was told me by an old Delsartean was useful to a man when embarrassed—this. [Shakes his hands violently.] Well, I have done it, but it has not restored the equilibrium of my brain, has not suggested any pertinent thought. I believe that these ladies and gentlemen, these fellow-alumni, brothers and sisters, must have profited exceedingly by the instructions of a certain Dr. Gunn on magnetism, and have absorbed my magnetism, or somehow taken my thoughts. I wish, however, to say a word about one matter, the recent organization of the Alumni Society. I welcome that most heartily, and sympathize with the statement that the vice-president made last night at the alumni dinner, that she was relieved now. She felt that you had undertaken the work of establishing upon a permanent basis the institute. She was relieved, and I feel relieved. Now, the thing that I would suggest at this moment is, not so much that you should have an object in life, which, as one of the students has just said, is one of the three grand requisites of success —you may remember the contented fellow who said that he was happy because he had health, he had a conscience free from any sense of ill-will toward his neighbor, he knew no enemy, and last, but not least, he had an object in life; for instance, he turned the grindstone down in the mill—but that you should keep in mind the *Phrenological Journal*. I dare to say that instead of being a private periodical, this magazine should be yours, the organ of the Alumni Association. I would have it the organ of the institute and the representative of the institute's men, speaking their thoughts, describing their work, and so letting the world know what they are doing for humanity. The grandest work which can be taken up—in connection, of course, with the work of the Divine Teacher—is the work you came here to learn about. Let the world know what you are doing for it, and as your agent or editor I shall be glad to help you to spread before the reading public what you are doing. I am sorry to say good-by. I have hardly begun to know you; you have hardly begun to know me. You have seen my better side, and you ought to stay a little while longer to see the other side and know just how I am rounded out. People love their friends most because of their weaknesses. The mother loves most her wandering son, her erring boy. Let us know more of each other hereafter, find out more of each other's human nature, and so come closer together in the bond of a common brotherhood and a common cause.

## ADDRESS BY PROF. NELSON SIZER.

It has been the custom for the faculty to speak first and the students afterward, and this year we have reversed the order, giving the field to the students and leaving the teachers to glean what may be left. As it was heretofore my lot to speak after all the other teachers, I have not been accustomed, therefore, to think of anything beforehand that I ought to say at the close. If I had it all my own way, if there was nobody else to talk, there is enough perhaps that I could find and ought to say that might be interesting. But the story of phrenology is a little like the piano, exceedingly rich; wherever you touch it, it sings, and therefore, fellow-students, as you go out into life you will learn to look on every man, woman, and child with interest; you will also learn that if you stand down nearly on a level with your congregation you can talk easier. I like to stand not much above an audience, and face to face when I am talking. When in former times the pulpit, like a swallow's nest, was put up fifteen feet high, I do not wonder at the seeming distance between preacher and people as their weighty words were filtered down through the wintry air of the churches where they did not have a fire. In New England, among the Berkshire hills in my day they carried to church a kind of stove for their feet, and still shivered through the long service.

When you go out into the field in the practice of phrenology a person will come to you and say, " I wish you to put me into the right pursuit, and what you tell me I am best fitted for I will do." You will feel responsibility as you never felt it before and you will invoke all you know of the subject. You may at first feel at a loss what to say, but when you have measured the head and studied the temperament, the character and talent as embodied in brain and temperament will dawn on you, and there will come a light, a glowing and an enthusiasm, which you will be conscious is derived from the human life under your hands. And you will find it out after a while, if you practise, that there is no human being, however insignificant he may look to you at first, that you cannot find something in him that is alive, that is human, that is humane. Occasionally a man sits under my hands when I am tired, the work of the week being nearly ended, when I have wished for an hour or two that nobody else would open the front door and come down to the examination-room to disturb my weary mind and body. In some such cases I sit and lay my hands lazily on his head, and begin to talk in a perfunctory kind of way, and all at once I find that the man is instinct with what makes humanity God-like and humane, and I spring to my feet and for twenty minutes perhaps I talk earnestly and rapidly, and when I get through I say to him, " Who are you, anyhow?" and he will tell me a story of experience, culture, suffering, triumph. He is perhaps, a great inventor who has done for mankind a world of service, and though he looks seedy, sad, and uncertain, and almost unworthy of your touch or talk, then I say to myself, " Well, I will never undervalue any human being as long as I live." Every man is worth saving, and sometimes men will pour out to you a story of their joys, of their sorrows or experiences that will almost make you ashamed of yourself, and yet you hardly wanted to touch them. You will have some wonderful transitions in ten minutes when you come in contact with people; and occasionally men will disguise themselves. Some years ago a clergyman was a member of the class; he boarded in a house with an Episcopalian clergyman and they had a chaffering debate about phrenology. The clergyman told our student, " Oh, that phrenology of yours is all bosh. They are pretty good judges of human nature and study the clothes, the make-up, and general appearance of a man, and form their judgments without any such thing as science connected with it." " Well," said the student, " you go in some time and try it and see; wear any clothes you please and see what he will say to you."

A few days afterward he went home from the class to his boarding-house and the clergyman accosted him. He said, " You have been and posted that phrenologist. I went over there disguised in a business man's suit—that is, a drummer's cutaway gray suit with a red neck-tie and all sorts of unclerical airs. I frowzed my hair down over my eyes and had looked as uncultured as possible, and I even used ungrammatical language. I went in as a kind of lout, acted awkwardly, and did not know anything worth mentioning, and he described my character with marvellous accuracy; but he was all the time talking to me as if I was a good deal better than I looked, that I ought to be ashamed of myself for not being more of a man than I was. Then he suddenly backed off and looked at me for a moment, and said he, 'Sir, you ought to have been educated; you should not have been left ignorant; you ought to have been a minister of the gospel of the grace of God.' I replied, 'What! me preach? Me be a minister—me?' 'Yes,' he said, '*you* ought to be a minister of the Word,' and insisted upon it, and I made up my mind he was sincere, or that you had been and posted him."

"How could I know that you would go, or if you did, what kind of a disguise you would put on? You might have gone in as a sailor."

There was a man in New York who had an examination once a year. He came once dressed as a gentleman, clean-shaven; the next year he came with a full beard; he came once as a sailor, in a lounging way, dressed in sailor's clothes, and each time had an examination written out under a different name. He came in once with overalls and lugging a carpenter's tool-box, and he asked the attendants near the front door, "Do you examine heads here?" and they said, "Yes." "Well, I guess I will have my head examined, but where will I put my kit of tools?" "Carry it right in and it will be safe." He sat down there as a carpenter, with his worn and dusty shoes and overalls, and I dictated the character and it was written out; and thus for eight successive years that man had examinations. At last he came and spread out the eight characters with the dates, and said, "There are eight characters that you have written out for me." I began to feel anxious. I can stand two or three, but to make it eight is rather heavy. I looked them over. "Well," said I, "what of it?" He replied, "I am one of the best friends of phrenology in this wide world. That line of written descriptions that you made for me a year apart have converted many and many a good man to phrenology. They would sit and listen to the different readings and see the similarity. There is one peculiarity about the matter. In the first examination I was told to increase a certain organ; in the next I was told to increase it; in the third I was told to increase it; in the fourth you said it was pretty fairly developed, and at the sixth and seventh it was spoken of as strong. I tried to cultivate it; I know I have cultivated it; and you have detected it all the way through."

A man who was a candidate for the mayoralty of New York, a German by birth, came in one day with a bundle wrapped in a newspaper and laid it on my desk. Said he, "I wish you would look at that." I opened it, and there was an old skull. Said he, "I wish you could tell me what you can of that head. They are making some excavations down-town and this skull was uncovered, and I told them if they would let me take it I would

carry it up and hear what the phrenologist would say about it."
I said of the skull, "In the first place, it is a German head; it is
of a man sixty or seventy years of age; I judge that by the teeth
and by the closed-up sutures of the skull. The owner of this
skull was an honest man, if there ever was one on the earth. He
was a high-tempered man, a very severe man, but he was just
and upright, and he was a healthy, hardy man; that is about all
I want to say about it. What of it? What do you think yourself—do you know anything about the man?" "Yes," he said,
"that is the skull of my father. I am repairing his tomb and I
thought I would bring this up."

My friends, you will get into corners; you must be on the alert;
never give up your watchfulness; never forget that you are God's
interpreters to men about things that are true. Study the subject under your hands sharply, and then have the courage of
your convictions in its utterance.

The last thought now is this, that you will be astonished and
abashed a hundred times at the amount of information which
people will suppose you possess. This is the most humiliating
fact in all our experience, that people suppose we know it all,
and I feel my weakness, but sometimes when I explain a man's
traits in such a way that he is astonished, and all his friends say
I have made it true to nature, why, then I pick up heart and
take courage; but I often feel weak, fearful, and solicitous when
I take hold of a head, as many times as I have had an opportunity of doing so. I trust to the indications of the science, and
it leads me to true results. Therefore do not be scared if, on the
presentation of a subject for examination, you do feel like saying,
"Who is sufficient for these things?" but in the fear of God and
love of man and of right go ahead, and the way will be open.
The ship does not go across the Atlantic without having to cut
water all the way and push, push, push; it is not easy work for
a ship to go through in five or six days; it is not easy for anybody
to examine heads. People say, "If I could only examine as easily
as you do!" My dear brother, I have been fifty years in practice and I work just as hard as I can every time. I venture to
say that it was just as hard work for the Apostle Paul to succeed
in his way as for the smallest man that ever tried to preach the
gospel. Paul had to do his best to preach like Paul, and the
rest of them, John for instance, would say, "Little children, love
one another;" it was the best and richest thing he could do.
Paul had to "reason of righteousness, temperance and judgment
to come," till "Felix trembled," and there is where moral agency
and responsibility come in. Each man has to put all the talent
he has into the work to fulfil his duty, and one can put all he
has into the work as well as another. Do your best first, last,
and always and you will constantly grow, and your proper success and your reward shall crown your efforts.

## DELIVERY OF DIPLOMAS.

It now becomes my pleasant duty, under the charter granted by the State of New York, to deliver to each of you the diploma of the institute for the class of 1890. For a quarter of a hundred years I have each year stood by this shining gateway and welcomed earnest students to the ripening fields which wait the gathering hand of industry and fidelity. My heart softens and my eyes moisten as with loving and tremulous tones I welcome each of you to accept this token of our fellowship. When in future years your eyes rest upon the seal of this document and you read the signatures, remember that our happiness in your behalf will be augmented in proportion as you realize our hopes in your success and usefulness.

---

## ADDRESS BY MR. ALBERT TURNER.

I would rather have a chance to say God-speed, in a meeting like this, than to say anything else. That is all I need to say, simply to wish you all success, and to reassure you, if necessary, of our good wishes, of our earnest interest, and our desire to back up your efforts. I will say this, though, that I congratulate this class on two things: first, of being this class and of being sure of the instruction this class has received, which may possibly be the last that any class ever will get of its kind. We know not what changes the future may bring us, and I congratulate the institute on having so good a class. The other thing I congratulate you on is the steps you have taken to make stronger and firmer and more effectual the phrenological bonds that ought to exist among all who love this subject. I refer to the formation of the Alumni Association. I believe that no step has been taken by any class that will compare in any way with the results that will come from the work you have started, and I have every confidence in the wisdom of those in this class and those of other classes that will unite with you in taking up the work in a new manner, with new strength and with new interest and new methods, and that the ball will start rolling after this year of 1890 in a way it never has gone before. I very often think when I am at my work, and I always have since the death of Mr. Wells, "What would Mr. Wells think of this? What would he think if he knew that we sold one hundred thousand copies of a certain book? What would he think of this plan or that plan of advertising?" I now am very glad that this step has been taken by the alumni, while those who are with us now have an opportunity of saying what they think of it and can give it their encouragement and counsel.

And now for the business part of phrenology. The Fowler & Wells Company is entirely separate from the Phrenological Institute in its interests financially, and while it is true that we work for money, it is also true that we do a great deal that does not mean money. The business interest of this subject has the missionary spirit, and whatever this Alumni Association may do to help the business interests of the house, it will aid the general cause, and we shall do all we can to help your progress and prosperity. Now, I simply wish for you all the success that you hope for, all the success that you honestly strive for, and we pledge ourselves to co-operate with you, and that we may do so please let us often hear from you.

---

## LETTER OF LEVI HUMMEL, OF THE CLASS OF '77, TO THE CLASS OF 1890.

"As a man thinketh in his heart, so *is* he," were the words of the wise man spoken nearly three thousand years ago. You will realize the truth of the above more and more as your phrenological investigations continue. You have taken up a wonderful and important science—*the* science of all sciences. You will find some noble and God-like "domes of thought," and others that are the dens wherein crawl the slimy serpents of envy, jealousy, hatred, and lust. Some that from their very nature think noble thoughts and are an honor to the race; others that dwell in the haunts of the propensities, the animal nature, and "their thoughts are of the earth earthy."

In some towns in your travels as phrenologists the right hand of fellowship will be extended to you by noble and generous souls, and you will feel at home in that town, and the people as well as yourself will feel that it was for the good of all that you were among them and you will feel loth to depart. In other towns you will "shake the dust from your feet" and regret that you ever got any of their dust on your feet to shake off. Such people will look on you with the cold eyes of suspicion. The ignorant, purse-proud, haughty will pass you by and the rest will look upon you as a sorcerer or fortune-teller. You will have bright and pleasant spots to dwell upon in retrospect and dark and dreary ones. But yield not to discouragement. You have undertaken a great and noble work. You can lift the fallen, comfort the sorrowing, correct the wayward, and guide the weak. You can read the mind of men by that science that solves the mysteries of man's nature that you have now been learning by attending the American Institute of Phrenology. What you now need is to bring your knowledge to the test of practice. You can feel proud of your profession; stand *erect* among the greatest. You can name with pride Dr. Gall, George Combe, Dr. Spurzheim, Vimont, Prof. Silliman, Hon. Horace Mann, Dr. Caldwell, Henry Ward Beecher, and many others, all of whom were an honor to the race and benefactors of mankind. You can do much good; help on many reforms. The mother will rely on you for counsel; her heart will tremble with hope or fear as you describe the noble qualities of mind and heart of her beloved son or daughter, or point out the shoals, quicksands, and breakers of passion that may wreck their young lives if not guided, corrected, and controlled. Yours is thus a noble and responsible work worthy of the greatest. Carry yourself as a true gentleman or lady in your profession. Be above suspicion. Amid discouragements look for better things, and when the tide turns be not too much elated. Cultivate equanimity of soul. Do all the good you can. Trust in God, and when the hair on your head turns to the color of the fleecy snow like that of your venerable teacher, Prof. Nelson Sizer, you can look back over a well-spent life, and listen with faith and hope for the call of the Master, "come up higher."

## NAMES OF THE CLASS OF 1890.

John C. Bateson, M.D., Pa.
Miss Pearle Battee, Ohio.
Mrs. Lillian J. Bausch, N. Y.
Samuel Brolin, Minnesota.
J. P. Burlew, Kentucky.
Neil Campbell, New Jersey.
Charles V. Carl, Canada.|
J. H. Chapman, Texas.
Anna Duval, Arkansas.
Wm. J. Duval, Arkansas.[1]
Miss Viola Evans, Indiana.
Wm. A. Fried, Canada.
Giles C. Gass, New York.
W. E. Hall, M.D., Texas.
Sarah Kirk Hare, New York.
Rev. S. R. Heebner, Pa.
Miss Ottillia C. Heine, Pa.
John J. Heins, New York.
Mrs. C. G. Henry, New York.
James Hickey, New York.
S. M. Hunter, England.

U. G. Hurley, Iowa.
J. T. Knott, New York.
Hugh Mackay, Brit. Col.
George MacDonald, New York.
James McCready, Virginia.
F. McNally, Iowa.
Mrs. Elsie E. McNally, Iowa.
Frank Mannion, Minnesota.
Jas. B. Moran, Pennsylvania.
L. G. Morgan, A.M., Canada.
B. A. Norman, Utah.
Cornelia C. Pattangall, N. J.
H. T. Phipps, Massachusetts.
H. W. Saunders, New York.
S. G. Sharples, Montana.
Victor G. Spencer, Iowa.
Sebastian Trawatha, Pa.
Anna V. Trawatha, Pa.
Richard Smith, Pennsylvania.
Guy T. Walter, Pennsylvania.
Wm. E. Wolfe, Indiana.

## LIST OF GRADUATES TO 1890.

WE are often written to by persons in different States to ascertain if "Prof. ——" is a graduate of the American Institute of Phrenology. Some persons whom we never before heard of have professed to be graduates of the institute, and even publish it on their circulars, endeavoring thus to secure consideration. The following list embraces the names of all the graduates up to and including the year 1890. All our students have a diploma, and would be happy to show it; and it would be safe to ask to see the diploma of those who claim to be graduates.

Abel, Miss Loretta, M.D., New York. 1877
Adams, Elijah, Missouri............ 1875
Adams, Miss F. R., Iowa.... ..... 1883
Ahrens, H. F., New York.......... 1888
Alderson, Matt. W., Mont. ...1875, 79, 80

Alexander, Arthur J., Indiana......1871
Alexander, W. G., Canada...........1884
Alger, Frank George, N. H..........1880
Anderson, Alex. H., Canada....... 1884
Anderson, Geo. W., Canada.........1897

Anderson, Samuel H., Pennsylvania.1867
Andre, James Wm., Pennsylvania ..1888
Arnold, Chas. H., Mass...............1870
Arthur, Willie P., New York ........1874
Asbell, B. F., Kansas ...............1889
Aspinwall, F. E., New York......1872-73
Austin, Eugene W., New York.......1878
Austin, Fred. H., Pennsylvania......1882
Ayer, Sewell P., Maine...........1868

Bacon, David F., N. H.............1875
Baker, Wm. W., Tennessee..........1876
Baillie, James L., Ohio......... .....1881
Ballou, Perry E., New York......1871-72
Bartholomew, H. S., Indiana........1885
Bateman, Luther C., Maine..........1871
Bateson, John C., M.D., Pa.......1880-90
Battee, Miss Pearle, Ohio........ .1890
Battey, O. F., Massachusetts........1883
Bausch, Albert, New York........1887-89
Bausch, Mrs. Lillian J., New York..1890
Beard, J. W., Virginia..........1887-88
Beahm, Rev. I. H. N., Virginia......1889
Beecher, Eugene, Connecticut.... .1870
Beverly, C. A., M.D., Illinois........1872
Beall, Edgar C., Ohio......... .....1877
Beer, John. New York. .............1878
*Bentley, Harriet W., Connecticut..1881
Bell, James. N. H... ...... .........1881
Boettger, G. W., New York.........1887
Bonine, Elias A., Pennsylvania......1868
Bonham, Elisha C., Illinois.........1875
Bousson, Miss O. M. T., New York.1877-82
Bradford, E. G., New York..........1888
Brady, J. Bradshaw, New York......1887
Brandenburg, C. W., New York.....1889
Brettel, Montague, Ohio...........1875
Brethour, E. J., Canada............1884
Brimble-Combe, Wm., Australia... 1886
Brolin, Samuel, Minnesota .......1890
Brownson, Rev. A. J., Indiana .....1884
Brush, Clinton F., New York.. ....1887
Bullard, J. H., New York..........1866
Buck, Marion F., New York........1868
Burlew, J. P., Kentucky. ..........1890
Brown, D. L., Iowa.... ...... ....1872
Brown, Robert I., New York........1887
Burnham, A. B., Wisconsin ........1881
Burr, Rev. W.K.,M.A.,Ph.D.,Canada.1884

Cady, Charles Everett, New York...1885
*Campbell, H. D. C., New York......1874
Campbell, D. H., Canada........1887-89
Campbell, Niel. New Jersey...... ..1890
Candee, E. E., N. Y....1873, 75, 78, 80, 88
Cannaday, J. C. B., Tennessee......1889
Carl, Chas. V., Canada.... .........1890
Carman, Lewis, New York........1883
Cassel, Harry K., Pennsylvania.....1886
Catlin, David C., Connecticut.......1877
Centerbar, J. S., New York....... ..1881
Chandler, G. E., M.D., Ohio.......1873
Chapman, J. H., Texas......1887, 88, 90
Chapman, May, Massachusetts......1879
Charles, G., Canada.................1876
Chesley, Egbert M., Nova Scotia.....1871
Chester, Arthur, New York.........1870
Clark, Perry, California........ ....1886

*Clark, Thomas, New Jersey........1874
Clarke, Rev. Jas. E., Maine.........1877
Collins, John, Wisconsin............1878
Condit, Hilyer, New Jersey..........1867
Constantine, Rev. A. A., New Jersey.1875
Constantine, Miss E., New Jersey.1875-84
Cook, J. R., Ohio..................1872
Corbion, William A., Pennsylvania..1888
Corfman, A. J., M.D., Ohio....... ..1886
Cowan, John, M.D., New York......1870
Cray, Edward A., Rhode Island....1885
Creamer, Edward S., New York.....1866
Crum, Rev. Amos, Illinois... .......1870
Curley, Miss Maggie, New York.....1887
Curren, Orville, Michigan...........1873
Curren, Thomas, Michigan.... .....1873
Curren, H. W., Michigan...........1874
Cutten, L. F., M.D., Canada. ......1888

Daly, Oliver Perry, Iowa. . ........1868
Danter, J. F., M.D., Canada........ 1870
Darling, Edgar A., New York. ....1885
Davidson, F. A., New York..... .1883-85
Davis, Edgar E., Iowa.............1885
Davis, Ida V., Wash..............1888
Davis, Wallace, Pennsylvania.......1875
Detwiler, E. W., Pennsylvania......1880
De Vore, S. V., Iowa..............1887
Dewing, W. F., Wisconsin..........1889
Dill, Rev. A. Cushing, New Jersey...1883
Diehm, Joseph, Kansas.............1885
Dieuis, R. O., Louisiana.............1890
Dodge, Lovell, Pennsylvania........1867
Dodds, Rev. D., M.D., Iowa.........1877
Doncaster, Wm. H., Pennsylvania...1888
Doolittle, Orrin, New York... .....1885
*Dornbach, H.F.A., Valparaiso, S.A.1885
*Downey, Rev. T. J., Ohio...........1867
Duncan, J. Ransom, Texas..........1875
Du Bois, D. C., Iowa.... ..........1877
Dutton, Geo. W., Nebraska.........1887
Drakeford, J. S., South Carolina....1889
Drury, Andrew A., Massachusetts...1882
Duval, Anna, Arkansas.............1890
Duval, W. J., Arkansas.......... 1889-90

Eadie, Andrew B., Canada..........1877
Earley, John, Ireland................1885
Ebersole, John P., Ohio.. ... .......1885
Eckhardt, P., Illinois....... ........1884
Emerick, B. E., Illinois.............1889
Emerick, Lycurgus, Illinois..........1876
Emery, C. Sumner, M.D., Ohio.....1887
Emery, Henry R., Ohio............1887
English, V. P., Lawyer, Kansas.... 1886
Espy, John Boyd, Pennsylvania.....1875
Estabrook, H. T., North Carolina...1889
Evans, Henry W., Pennsylvania.....1867
Evans, Miss Viola, Indiana.........1890

Fager, Andrew C., Ohio...... . ....1887
*Fairbanks, C. B., New York........1872
Fairfield, John C., Pennsylvania ....1876
Fariss, F. A., Virginia............1885-87
Fawcett, W. P., Virginia...........1889
Ferry, A. L., Illinois.............1881-84
Field, J. H., Colorado............ ...1866
*Fitzgerald, Miss D. W., New York..1887

* Deceased.

* Deceased.

Lischer, M. E., New York...........1883
Lockard, E. M., Pa.....  ....1883, 1884
Loomis, Benj. F., California......  ..1886
Lomison, Wm. A., Pa......   .....1886
Luxford, F. Wm., New York........1887
Lyon, Chas. B., Michigan...........1880

Macduff, Rev. R. E., Kentucky......1872
MacGregor, Alex., New Jersey......1888
Mack, II. Q., New York.............1867
Mackenzie, J. H., Minnesota........1873
Macrea, Miss Flora, Australia......1884
Maxwell, Robert G., N. C.........1887
McCoy, Jason B., Ohio.............1885
Mackay, Hugh, Brit. Col...........1890
McDonald, Duncan, Mich......1867, 1882
MacDonald, Geo., New York ....... 1890
McIntosh, James, Ohio..............1867
McDavid, J. Q., S. C...............1874
McNeil, James, New York...........1873
McCrea, James, Illinois............1873
McCready, James, Virginia..........1890
McFaden, Rev. J. D., Phil. Pa.. ....1889
McGuire, C. F. M., New York.......1888
McKee, William C., Ohio...........1879
McKenna, Thomas, R. I..  .........1888
McKinnon, C. B., Canada...........1889
McKim, John J., Mass..............1887
McLaughlin, Canada....  .... ....1882
McNally, F., Iowa....  ...........1890
McNally, Elsie E., Iowa............1890
McNaughton, S. S., New York.......1871
Mann, H., Jr., Vermont.....  .....1883
*Manners, J. H., New Zealand......1877
Mannion, Frank, Iowa...  . . .1879, 1890
Martin, Edwin E., New York........1880
Matley, John, California...........1872
Matlack, A. S., Ohio  .............1875
Mason, James, Mass................1880
Mason, Lott, M.D., Illinois.........1869
Mason, A. Wallace, Canada.........1874
Masters, Edward, Australia.........1888
Mehan, F. I., Mich................1889
Merrifield, John C., Canada.. ......1868
Meller, Frank J., Illinois..........1881
Memminger, T. F., West Va.........1881
Meyer, Robert C. J., Illinois........1888
Mills, Joseph, Ohio . ............1880
Mills, Rev. J. S., Ohio.............1872
Michael, J., Minnesota.............1868
Miller, Rev. E. A., Virginia. .......1889
Miller, Mrs. E. A., Virginia........1889
Miller, B. Frank, California........1882
Miller, E. P., M.D., New York.. ....1866
Miller, Henry, Michigan...........1887
Miller, John C., Ohio....  .........1888
Moatz, Lewis, Ohio.............. 1869
Moore, Joseph II., N. C............1877
Moran, Jas. B., Penn...............1890
Moran, Maggie L., New Jersey......1885
Morgan, L. G., M.A., Canada. ....1890
Morrison, Edward J., Illinois.......1868
Morris, Geo., Oregon.........1878, 84, 88
Morris, Marietta M., Oregon........1888
Mully, A. E. F., New York....  .....1882
Musgrove, Wm., England............1875

Newman, A. A., Illinois .....  .....1867

Nichols, Perry L., Iowa.............1887
Norman, B. A., Utah  ..........  ...1890

Oestergard, J. C., Denmark....  ....1883
*Oliver, Dr. F. W., Iowa.............1885
Olney, Henry J., Michigan ..........1875
Orvis, Heil F., Wis.............1886, 1887
Osgood, Rev. Joel, Ohio............1882

Pallister, Wm., Canada..  ........  ..1882
Parker, R. G., Missouri............1874
Parker, Howell B., Ga...  .....1875, 80, 85
Pattangall, Cornelia C., New Jersey.1890
Patton, Edward M., Illinois.........1874
Patten, Wm. Perry, Nebraska.  ....1876
Patterson, John A., Missouri........1870
Paulsen, John II., La..............1877
Pentney, John, Canada....  .......1877
Perkins, Fred. W., Missouri.........1889
Perkins, Mary A., Missouri.........1889
Piersoll, Sampson II., W. Va..  .....1870
*Perrin, Edward M., Kansas.........1869
Perry, A. D., Mass...  ...........1883
Petry, Daniel F., New York........1866
Philbrick, S. F., Ohio. .........1873, 1874
Phipps, Henry T., Mass......1887, 1890
Pooler, Mrs. F. M., Mass. ...,......1887
Potter, Miss Helen, New York......1887
Pierce, David F., Conn.............1868
Powell, L. M., M.D., New York.....1886
Pratt, Benj. F., M.D.. Ohio.........1875
Prather, Miss M. O., Kansas........1876
Price, David R., Iowa......  ..... 1868
Purcell, E. M., Iowa...............1874

Ream, Elmer, Indiana..............1885
Reed, Anson A., Conn.............1868
Riddel, Newton N., Nebraska.......1887
Rhone, Geo. W., Pa....  ..........1886
Richardson, M. T., New York.. .....1870
Richards, William, Pa..............1873
Righter, M. Helen, Illinois..........1876
Richie, Porter D., Illinois...........1871
Ribero, Manuel, Spain..............1887
Robbins, T. L., Mass ....  .........1872
Roberts, I. L., Florida.............1872
Roberts, Jas. Thos., California......1882
Roberts, Margaret E., Pa ....  ....1882
Robinson, Frank O., Tenn..........1885
Robinson, G. M., Illinois...........1881
Roeseler, John S., Wisconsin...  ....1884
Rogers, Ralph, Tenn...............1875
Romie, Paul T., California. .. ......1877
Rosenbaum, F. Wm., Ohio..........1878
Russell, Geo. P., Tenn..  ..........1888

Sanches, Mrs. Marie, Sweden........1880
Sargent, C. E., N. II..............1874
Saunders, H. W., New York....  ....1890
Scheaffer, J. S., Iowa..............1884
Scott, Martha A., Colorado.......  1881
Scott, Rev. Wm. R., Illinois.........1883
Senior, F. D., New York. ...........1872
Seybold, F. J., Illinois.............1870
Shamberger, Daniel, Virginia.......1885
Sharples, S. G., Montana...  .......1890
Shultz, R. C., M.D., Iowa...........1876
Sievert, Miss Sophie, New York,.,.,1880

* Deceased,

* Deceased.

## TO THE MILLIONAIRES OF AMERICA.

### From a Graduate.

I WISH to say a word which may call forth to action the benevolence, the conscientiousness, and the hope now passively resting and waiting for an opportunity to bestow a benefaction that will keep pace with the progress of civilization till every heart and home of man shall feel its benign influence in all the ages.

The American Institute of Phrenology, incorporated by special act of the Legislature of the State of New York in 1866, has done and is doing work that will command the admiration of philanthropists of all states and conditions, and although it has had to struggle with the opposition of prejudice and the lack of ample means, it has not only held its own in maintaining the science and scheme of human progress which Gall and Spurzheim and others had begun and formulated, but has elaborated and extended their teachings. The Alpha and Omega of this institution is the science of mind and body studied *together* in their *mutual* relations, and it is for the practical application of this grand scheme that I have ventured to write this appeal.

The world of society, education, and business is gradually feeling the blessed effects of the reforms which have been and are now being promulgated in this school. Its literature, which is now assigned a department in nearly every public library and has become a household word in thousands of homes, has had a tendency not only to influence public laws and public opinion, but has struck deep to the roots of the medical science as regards personal health, sanitary improvements, and hygienic diet. It has expounded the harmony of physical functions with those of the mental, and a vast range of kindred subjects which may be grouped under the head of the "science of human improvement." As a result of its teachings thousands are encouraged and strengthened to perseverance in self-culture in all its phases, are assisted in overcoming the defects from bias of organization, and are redeemed from the thraldom of many vices of heritage or acquirement, and men and women are directed *individually* to their *right places in life.* My venerable preceptor, Professor Nelson Sizer, has made some remarks on this subject which are so pertinent and practical that I will take the liberty of quoting them verbatim :

" The great mass of the people, perhaps ninety-five in the hundred, are obliged to earn their bread, and consequently the question ' In what way can I best earn my living?' must be met and answered, wisely or otherwise, by the majority of persons. . . . 'Business,' which is the general name applied to the channels in which men struggle for success, must take a prominent place in human thought, and men are anxious to know what they may best do in order to secure the success requisite to their proper support and for the best achievement possible in life. As pursuits are of varied character, requiring different kinds of skill, talent, judgment, facility, and industry, and as men also vary as to capacity and skill, it is a matter of high importance that each person should find the pursuit which is best adapted to meet his talents and characteristics. It will appear certain that if men adapted by talent or constitution to particular lines of effort shall be placed where the duties are not adapted to their constitutional qualities, failure and disappointment will be the result.

It may seem a startling statement that men vary as much in
their skill and talent in respect to different pursuits as the tools
of a trade vary in fashion and quality, but we think this to be
the case. . . .

"To one who knows how to study human character and to
observe and understand the qualities requisite for different pursuits, the awkward and illy-adapted efforts which some people
make to secure success in the different kinds of business awaken
in him a wonderful interest and present a problem the successful solving of which would be a blessing to the world, and must
challenge its attention. They use wrong tools awkwardly and
badly do their work. It makes one think of a man trying to
bore a hole with a screw-driver, or to cut off a board by boring a
line of auger-holes across it. If every boy and girl could be in-
troduced to the line of industrial and economic effort to which
he or she were, on the whole, best adapted, it would double the
prosperity and material good of the next generation and greatly
enhance the happiness of the race, besides abolishing poverty
and nearly abolishing crime. As the choice of a pursuit is in a
great degree the foundation of a man's fortune and happiness in
this world, any aid obtainable in regard to the right use of one's
faculties in the right direction deserves and ought to command
thoughtful attention. If one wastes his season of apprenticeship,
the boyhood and early manhood of his time, on a pursuit to
which he is not by temperament and mental development adapted,
and does not find it out until he has wasted years of precious
time, it must be a damper on his whole life."

These remarks require no comment—they are the thoughts of
one who has observed and studied human character more closely
and done more to put people in their right places in life than
any other man in this age or century. The object of this letter
is to enlist the sympathy and means of some one to help us to
carry on the work on the plan which has been laid out and followed, and to extend our borders by making it a *free* public institution instead of a private enterprise. We have ample material as a foundation, and it is strong and wide enough for a
great school, but we have not the money as yet of a Peter Cooper,
a Cornell, or a Vanderbilt, to erect a building that will stand as
a monument and a light in the centuries to come.

Concerning the Institute of Phrenology as an incorporate body,
I will simply refer to its charter, which is always published in its
annual announcement, in which will also be found various items

of interest and information concerning it. I will add, however, that its graduates have now formed themselves into an alumni association, which has for its aim and object the fostering of its alma mater, the enlargement of its scope and usefulness, and the securing of a permanent and suitable structure for the uses of the institute. But as our ranks are composed of people of moderate *financial* means—principally clergymen, doctors, school-teachers, and educated artisans—our progress must necessarily be slow unless it is hastened by the wealth of some noble giver.

We want—First, a building in a central place in New York City that will be accessible to all who reside in or who visit the metropolis. The building should be of suitable dimensions to contain lecture-halls, class-rooms, and fire-proof rooms for the preservation, use, and exhibition of the libraries and cabinets of crania, busts, casts, portraits, etc., and additions which will be made from time to time to its present valuable collection. If such a building were donated to the institute, I think it is safe to say that it will appropriately bear the name of the donor,

Second, we want to establish the several professorships by a permanent fund that will place the collegiate branch of the institute on a basis of financial security and comfort.

Third, we want to establish a "Bureau of Utility," in which all persons can consult and have the advice of the best and most experienced phrenologists in the world, and can have full written reports or delineations of character, with specific directions how and what to do to make the most of the inborn capacities and virtues they possess; to show each in what business or pursuit he will best succeed, thereby inspiring to self-knowledge and self-culture; also to disclose to parents their children's natural callings, dispositions, defects, and the mode of government best adapted to each. Employers of every description of labor would soon discover the practical advantages of applying to such a bureau for the help most suited to their several departments, which can be made an invaluable instrumentality for personal development, improvement, and happiness, as it will furnish the very best means for building up and strengthening the noblest elements of manhood to be the factors in our religious, social, and business life. If this bureau were sufficiently endowed, all this could be done free of charge to all comers. Such an institution properly equipped would become a great and unequalled benefaction.

May God inspire those possessed of means to thus reach out a

3

helping hand to give their fellow-mortals more light—more happiness. This is a work which applies to all the varied interests of humanity—a work calculated to inspire and lead to a higher, truer, and nobler human life.

To him who loves his fellow-man the best,

I am, ever an obedient and humble servant,

GEO. MACDONALD.

## ALUMNI ASSOCIATION.

THE class of 1890 united with the "American Phrenological Society" for the purpose of discussing the expediency of organizing an association of the alumni of the American Institute of Phrenology. The meeting, held in New York October 1st, 1890, was called to order by George MacDonald, who was chairman of the committee of the class 1890.

Mr. MacDonald was then elected chairman to preside over this meeting. Pearle Battee, class 1890, elected secretary.

Proposed for discussion : The organization of the graduates of the American Institute of Phrenology into a permanent association; also the providing of a home for phrenology; also the feasibility and practicability of holding a convention of phrenologists at Chicago during the World's Fair.

Remarks were made by Mr. Sullivan, class '85; Dr. Bateson, '90; R. Smith, '90; and Frank Mannion, '79.

Reading of resolutions and programme made by the committee chosen September 23d by the class of '90. Dr. Hall, '90, earnestly sustained the resolutions.

Mr. Sullivan moved that this meeting be the first meeting of the alumni of the American Institute of Phrenology. Seconded and carried.

Prof. Sizer spoke in favor of an alumni society and offering to give the alumni space in the *Phrenological Journal* for the publication of transactions and matters of interest to the association, if the alumni will furnish the matter to fill it.

Remarks by Mr. Mannion, MacNally, '90, the chairman, and Dr. Bateson.

Question called for and carried.

It was moved and seconded that the present chairman and secretary act temporarily for the alumni. Carried.

Moved and seconded that the chairman appoint a committee of three besides himself and the secretary to frame a constitution and by-laws. Carried.

Moved that Prof. H. S. Drayton, Frank Mannion, and Guy Walters, '90, be that committee. Carried.

Then followed the signing of the names of members of the alumni.

Adjourned to meet Monday, October 6th, at 2 P.M.

PEARLE BATTEE, Secretary.

On the 6th of October, the students in attendance at the Institute of Phrenology organized an Alumni Association. The constitution and by-laws herewith set forth received unanimous approval, and Professor Nelson Sizer was elected president. With a graduate list of five hundred, the movement inaugurated by the class of 1890 seems entirely expedient. The wonder is, indeed, that such an association had not been organized before. That the feeling of fraternity which is born of prolonged association should be one of the inspiring elements that prompted such a conclusion goes without saying; but other and higher purposes are involved, as appears in the following copy of the instrument of organization:

## CONSTITUTION OF THE ALUMNI OF THE AMERICAN INSTITUTE OF PHRENOLOGY.

### ARTICLE I.

Section 1. The undersigned graduates of the American Institute of Phrenology, appreciating the importance of organization for our mutual improvement and interests and for the advancement and protection of the science of phrenology throughout the world, have associated ourselves together under the name of the Alumni Association of the American Institute of Phrenology.

Section 2. The objects of this association are the collecting and preserving of historical and scientific data pertaining to phrenology and its kindred sciences; the collection and preservation of skulls, casts, drawings, charts, and scientific apparatus pertaining to the same; and for the purpose of contributing essays and articles for discussion at its meetings and for publication; and for the purchasing of real estate for a suitable and permanent home for said collections and the business of the association, and the collection of funds for these and such other purposes as the association shall deem proper.

### ARTICLE II.

Section 1. The officers of this association shall consist of a president, first vice-president and other vice-presidents, secretary, treasurer, and a committee of three members on ways and means, all of whom shall be elected annually.

Section 2. All meetings of the executive board of the association shall be held in New York City, or at such other place as they in their discretion may select.

Section 3. The president, first vice-president, secretary, treasurer, and committee on ways and means shall constitute the executive board of the association, and the members of said executive board shall reside within a radius of two hundred and fifty miles of New York City.

Section 4. The association shall admit to membership the members of the faculty of the American Institute of Phrenology, and any graduate of said institute, of good moral character, whose application has been approved by a member of the association; the elections to membership shall occur subsequent to the presentation of the applicant's name at any meeting of the association.

Section 5. When a person has been elected to membership he shall assume such membership as soon as may be convenient, sign the constitution and by-laws, and pay to the treasurer the sum of one dollar as initiation fee.

Section 6. Five members of the executive board shall constitute a quorum for the prosecution of business.

Section 7. The annual dues shall be one dollar for each member. The executive board shall have power to levy such assessments as may be deemed necessary to carry into effect the objects for which the association has been organized; but no assessment shall exceed the sum of five dollars per member per annum.

Section 8. The annual meeting of the association shall be held at such time and place as the executive board shall designate. The executive board is authorized to call a special meeting of the association by mail, stating purpose, at any time it may be deemed necessary; and said executive board shall notify each member of the association, by mail, of time and place of said meeting, not less than four weeks prior to holding the same; and twenty members at said meeting shall constitute a quorum.

Section 9. All members not present at annual meeting may vote by proxy, provided said proxy is authorized by letter from said member to proxy, and said letter shall be delivered to and retained by the secretary of the association for one year.

Section 10. This constitution or any clause thereof may be altered or amended at any annual meeting of the association by special resolution, provided that a copy of said resolution *shall* be sent by mail to each member of the association not less than four weeks prior to said annual meeting, and shall require for its approval and adoption not less than two-thirds of all the members present and two-thirds of those represented by proxy.

## BY-LAWS.

Section 1. The election of officers shall take place on the date and at the place designated by the executive board, and by the ballot of those present and by proxy of those absent, the majority electing; should a tie occur, the presiding officer shall be authorized to give a casting vote.

Section 2. It shall be the duty of the president (or in his absence the first vice-president) to preside at each meeting, preserve order, and regulate the debates.

Section 3. It shall be the duty of the secretary to keep a clear record of the proceedings of each meeting, to read the proceedings of the previous meeting, to give notice of all meetings by mail, and to preserve and keep all records and documents belonging to the association. It shall also be the duty of the secretary to write and answer letters and communications in behalf of the association.

Section 4. It shall be the duty of the treasurer to keep a regular and correct account of the monetary matters of the association, to collect all moneys or dues owing, from members or otherwise, and to pay all orders countersigned by the president or a member of the executive board. It shall also be his duty on the expiration of his term of office to present a written record of his doings in his official capacity.

Section 5. Any member who shall fail to pay the annual dues for two successive years shall, at the discretion of the executive board, be suspended from the privileges of membership.

Section 6. It shall be the duty of the executive board to exercise a general supervision of the literary and business affairs of the association, to advise with regard to the character and cost of books, crania, casts, busts, portraits, and other scientific properties which it may be deemed desirable for the association to possess, and to suggest topics for discussion at the meetings. It shall also be the duty of said board to provide a suitable room or accommodations for the use of the association at its meetings.

Section 7. These by-laws may be altered or amended by a vote of two-thirds of all the members present, at a general annual meeting of the association, and two-thirds of those represented by proxy.

## SUPPER OF THE CLASS OF 1890.

On the evening of the 9th of October the students of the American Institute of Phrenology, class of 1890, members of the board of trustees, several of the graduates and invited guests, numbering in all about sixty, sat down to a well-appointed table. Prof. Nelson Sizer, president of the institute, acted as master of ceremonies, and having rapped for order, the Rev. Samuel K. Heebner, class of '90, offered an appropriate invocation. After the bountiful repast had been enjoyed, President Sizer opened the intellectual programme by saying:

"The American Institute of Phrenology was incorporated in 1866. In that year its first class was taught in New York, and consisted of six members. The second class, that of 1867, had fourteen students, several of whom were connected with professional work in other departments, such as teaching and literature. One of the students of that second class has become widely known as a successful lecturer on phrenology in the western country, and has made in his profession a handsome fortune. The institute graduated in its first 25 years 487 students, an average of 19½ per year.

"Yale College, from its commencement in 1701, during its first twenty-four years graduated 141 students, or an average of less than 6 per year, and we, during our twenty-five years, have graduated 487 students, or over 19 a year.

"Four years more and we shall celebrate the centennial of Dr. Gall's first public lectures on phrenology. In August last the sixty-second anniversary of his death occurred; in one month from to-morrow we shall celebrate the translation to a brighter sphere of Dr. Gall's first beloved disciple, Dr. Spurzheim, whose body Boston laid, as the first occupant, in Mount Auburn, her beautiful city of the dead, in the year 1832—fifty-eight years ago. These were the founders of phrenology. The best citizens of Paris and of Boston respectively gave the strangers and benefactors peculiar honors at their graves, regarding them as among 'the few, the immortal names that were not born to die.'

"We are making history. Monday, the 6th of October, this Association of the Alumni of the American Institute of Phrenology was formed. We are the constituent members. May our honored successors cherish the pleasant work and field we now open to them.

"We trust that our successors shall celebrate the centennial of this institute in this imperial city, and that its alumni will be numbered by thousands and grace every eminent position in the world of letters, law, divinity, science, and business. The study and culture of man in mind and morals are the objects of our effort, and time and truth shall reward it."

The secretary, Dr. Drayton, then read letters of regret and congratulation from Dr. H. A. Buttolph, L. A. Roberts, E. W. Austin, Prof. Howell B. Parker, of Georgia, Dr. A. H. Laidlaw, Dr. R. Shultz, and Prof. S. S. Packard, after which Mrs. Charlotte Fowler Wells, the vice-president, responded to the toast "For the Future."

"Our president says that he wishes me to speak something of the future; of course he must mean a prophecy. I prophesy that phrenology is to be known the world over; that it will be taught in all schools; that children will have their heads examined early and be put in the right direction, and when they start right they will not deviate from that path. Phrenology is bound to succeed, for it is true, and truth must prevail; it is one of God's truths. We may not live to see the bright star that will come eventually, but we can see the forerunner of the star. . . . The oak is an example; so perhaps the slow growth of phrenology indicates that it is to last as long as man lasts. I congratulate the Alumni Association of the American Institute of Phrenology that has just been organized, and I prophesy for it greater attainments and hope it may do wonders. There is a great deal for you all to do. I wish to express to you my thankfulness. I would if I could, but I have not the power to express to you how thankful I am for the relief you have given me in the stand you have taken, the efforts made and the promise given to work for the prosperity of the American Institute of Phrenology, and securing of a building of its own in which to do its work and care for its valuable cabinet of busts, casts, crania, portraits, library, and illustrations. I have felt that I had a load upon my shoulders; you have removed it. I felt that when the home for phrenology was obtained I could die easily. The alumni have taken from my shoulders what I had felt that I must do myself, and I thank them for it, and I hope they may do more than they expect."

The next toast was "The Alumni Association," to which Mr. George MacDonald, of Albany, class of '90, said:

"Mr. President, ladies and gentlemen: It gives me great pleasure to listen to the sentiments expressed. I feel my incompetence to address you with regard to the work laid out by the alumni; it is work of the greatest importance to this institute and to the world at large. We have stated in our new constitution that we shall gather funds for the purpose of building a home for phrenology. We have heretofore, as phrenologists, been empty-handed, so to speak, with regard to a depository for all the precious relics of phrenology. Our honored 'mother,' Mrs. Wells, has had her heart set on that subject for many years, and I trust she will live to see the day when the foundation stone of the universal home of phrenology will be laid. . . . While we are here to-night, my friends, let us resolve that we will not leave a stone unturned until we have laid the foundation stone, and if it cost a million dollars we shall carry it through. We have the material here, we have the will here, we have got the intellect here, we have got the power to influence others to come in and fill the fund that will raise a monument to those who have worked for the good of the human race. My friends, let us pledge ourselves as men and women to accomplish this object."

The president then called on Mr. Matt Alderson, late editor of the *Avant Courier*, Bozeman, Mont., to respond to the toast "Our Graduates." Mr. Alderson said, in part:

"Mr. President, ladies and gentlemen: I feel that it is no small honor to speak for the former students of this institution. Professor Sizer has called your attention to the fact that they number nearly five hundred, and I think that there are few institutions that have been attended by individuals from parts of the world so far distant. I remember, when I attended in 1875, that it took me about thirty days to get here. I travelled about three thousand miles; and there were students from Canada. At other sessions there have been students here from the Old World, from Australia, and there are some present from the Old World.

"Many who have attended the institution have attended it solely for the purpose of acquiring information concerning human nature, not for the purpose of practising the science as a profession.

"I think that it would be the unanimous voice of all persons who have attended the American Institute of Phrenology that the information they obtained there has been invaluable to them. They find use for a knowledge of human nature in every walk of life, and whether they use it as a profession or not, they make use of it in the editor's chair, on the bench, in business offices, at home, everywhere. I am pleased to meet so many representative people at a gathering of this kind. I certainly think it is a creditable gathering, and all here should go out satisfied that phrenology is making rapid advance in the opinion of the world, and is being more generally accepted."

The succeeding toast was "The Trustees," and Dr. Drayton, secretary of the board, was invited to speak on their behalf, which he did in part as follows:

"My friends: As one of the trustees, I am expected to be able to say a word in response to the toast just propounded. There are five of us; two are absent to-night; the president of the board sits on my immediate right; the vice-president of the board sits also on my right, next the president. Dr. H. A. Buttolph is also a trustee, a gentleman of wide experience, a specialist in medicine for the insane, unable to be here to night, but always expressing himself as most warmly in co-operation with us. Mr. L. A. Roberts, from whom we have a letter, is the other trustee.

"From the very beginning, so far as I know, this board has been actuated by but one purpose, and that is to make the Institute of Phrenology a working, educational, useful organization. I remember in the early days of my connection with the institute that the classes that assembled were not remarkable for numbers, nor were the lectures remarkable for numbers, and yet the student fee of one hundred dollars a course was deemed by no means excessive. Those who came paid the money freely, and went away perfectly satisfied with what they received for it. I believe that a score or so of lectures were delivered at the first, and the number was gradually increased. There were but a brace of lecturers at first, and those lectures were mainly delivered by Mr. Samuel R. Wells, then a trustee, and Prof. Nelson Sizer.

"The trustees have not been backward in making use of their opportunities and the cash that fell into their treasury, and therefore no large fund has been accumulated. So the trustees —and I speak without thought of personal credit in the matter —have always worked for the growth and benefit of the students, not for the institute merely. Of course they have had their expectations; they have thought a great deal all these years, as one of them has intimated just now, with regard to erecting a permanent building, the creation of some granite block, that would fitly represent their aim, but theirs has been the spirit of liberality for the student, not of economy for the sake of creating a large fund. There have been those that have proffered financial assistance. We have been offered thousands of dollars, but somehow or other the money has not come. There are promises still in the air, and perhaps they will be realized ere long, and with this growing potentiality of an Alumni Association at hand very likely the funds will come along for a start of the structure so warmly depicted by Mr. MacDonald, and then the Institute of Phrenology will be established on a solid basis, with its own lecture halls, and its own museum, and other facilities much needed to perfect, in grand review, the system of instruction. But even now, as I look over this large company, it seems that we might be strong enough to make a start in the direction of our purpose.

"And then those statistics which you have heard from the president; they are very encouraging. It makes a trustee feel that he can go on; that he can spend all the money that may be placed in the treasury, and more will come. In twenty-five years, five hundred students! Compare the record of Williams and Yale and Princeton, and some other colleges that are leading to-day; look over their record of twenty-five years from the beginning. We have done well; we expect to do better; and with your co-operation, friends of the Alumni Association, so kindly, so earnestly, so zealously promised, I am sure we can go on and do grander things."

"Our Class" was then announced, and Mr. Frank Mannion said:

"I am sorry that an abler person has not been selected to speak for the class, for I feel unequal to the task; but with the inspiration from so many fair women and brave men, I may hope to say a word or two in behalf of the class. I may say that those before me have outlined our work. Our president has told us that we are making history. Do we realize that? Can we say to our president that we do here realize that we are making history in the sense that we represent the phrenological world? There are many different kinds of worlds now; the world of science, the world of ethics, the religious world, and so on—so that this school and this place fairly represent the phrenological world. The eyes of the phrenological world, we may say, are upon us as truly as the eyes of Europe were on the army of Napoleon ninety years ago. Great things are expected of us; a great charge is placed on our hands as a part of the alumni. It is the erection of the building that is to be the memorial of the work that has been done in the past, and a foundation for the great work before us in the future.

"I thank you in behalf of the class for the privileges we have enjoyed. I thank you in behalf of the class for the promises of earnest interest in the future; and we will strive to build up this great work that is before us, this great institution. We have signed our names to the constitution that calls on us, as honorable men and women and as classmates, to work and labor in that behalf, and we are pledged to the support of that constitution. We are here to-night and perhaps will never all meet again, for we can never again enjoy the pleasures of to-night. Let us go out as did those from the Last Supper of history, to preach the gospel of phrenology to all the world."

"Our Guests" found in Mr. W. H. Vanderbilt, of Brooklyn, an advocate, and in speaking of them he said:

"Mr. President and friends: I am a guest here this evening. I have not been a student yet, but have been a reader of the books for some thirty years, and have taken great interest in the subject all that time, as my business would permit. Some of the things I might have said have been outlined by the former speakers, and particularly by the speaker for the former students. I feel that the whole community is interested in this work, parents, teachers, and business men. I believe the world is interested, and the world has contributed well to the classes of this institute. Bacon has said that reading makes a full man. I have been a reader of the books, but have hoped at various openings of the sessions that I could be able to come into the class, but the opportunity has not been given me. However, I may send one of my children as a substitute later, and then I shall be just as well pleased. I feel that whatever we can do for those that are about us will do us as much good as if it were done for ourselves.

"I am glad to feel that the class express themselves as ready to go out and work for the institute, and in my reading of addresses by former classes, the same sentiments seemed to prevail. I believe that the students, coming from scattered portions, will make that growth assured."

"The Ladies" proved a sentiment that Prof. Charles E. Cady, class of '85, promptly rose to greet, saying, in part:

"Mr. President, ladies and gentlemen: What should be said for the ladies? I am a woman's-rights man, whatever the word implies. I have been a teacher a great many years in public and private schools; have tried to teach even those hieroglyphics that our friend [referring to the reporter] is putting down as I utter the words.

"I should like to relate something of two incidents that illustrate the value of a word dropped in season. A pupil of mine, years ago, determined to study medicine. She studied it under great difficulties. First her pastor censured her, because it was not proper; it was not modest for a woman to study medicine. Then another friend said, 'Well, if you are going to study medicine, you can walk on the other side of the street;' and a word which I dropped, she says, encouraged her more than anything else to go on. She is one of the smartest women in that line I know. In this city there is to-day one of the most successful lady stenographers, whom I am happy to call one of my pupils, who now teaches it, and is employing eight or ten others to assist, and she has a contract for doing some work that will bring her a thousand dollars or more. She left my school, went to work for an employer, left him three or four times, and at last said she had left him for good. 'Of course you have,' I said; 'you have done that three or four times, I believe just exactly what you say.' 'No, I have left him entirely this time, and have come to ask your advice about opening an office.' I advised her to open an office and begin work, and she has done it and won success.

"I remember when it was hardly respectable for a woman to do anything in the way of self-support unless it was school-teaching. Why, the people would have looked with holy horror to see women in the stores—business women, as we see them now. The world has not exactly grown out of that idea yet.

"Phrenology opens a good field for one of the best-paying occupations that a woman can take up. I am delighted to see women studying phrenology. I believe that they may make better examinations than men. They have the deftness of finger, the delicacy of touch which should enable them to examine better than most men can. If they have intuitions, as women are credited with having, that should help them in reading character. Now, ladies, I am delighted to see you in the study of mental science. As a teacher and an observer and reader, I may say this, that there is a great field for women in phrenology, and I am delighted to see so many engaged in that study.

"I ought to compliment the phrenological school on what it is doing in this line, having ten lady students in this class and thirty-two gentlemen; but that not being my topic I must not touch on it. I will simply say that I hope that the classes, instead of being as my class was, five years ago, composed of one woman and forty or fifty men, the time will come when those conditions will be equalized."

Mr. Albert Turner was called upon and spoke for "The Field," saying:

"I certainly have a better excuse than Mr. Cady or anybody else who pleaded the lateness of the hour, and the fact that others before them had said what they might have said.

"The field that is before the phrenologist is a good subject, perhaps, for me to consider. I stand on a high point in this matter. I can view it from many ways, and think I know as much about it as you are likely to have a chance to know. There is scarcely a week that we do not get letters from somebody wanting a lecturer, and every mail brings letters asking for information on our subject. The writers want a list of books, or want to know about examinations or something else that is kindred. It is not always that which will bring us money, it is not always business, but the people want information and expect answers to their questions. Now, that being the case, is it not fair to presume there is a good field for you, an open field? The phrenologist who gives advice for pay does very much more service than is paid for. I have studied this ground carefully, and I have taken it very strongly that there is no amount of money that it is possible for you to collect from anybody who may consult you that will form a full equivalent for your advice.

"Now, then, if you want to make money, if you look upon this as a business enterprise, and you are prepared to do it, if you have the qualifications, if you have the capital either in brain or bank account, and will put yourself in the right relation to the

public, you can make money.  Others have done it, and many of
them.   I am speaking now to those who are planning to take up
this field, to fill it, to do justice to it, to go in with a spirit that
will not be discouraged.   If you are honest, square, earnest, the
people will pay you.   I hope that this class will do their best to
occupy the field.   If at first you cannot lecture, examine heads;
if you cannot examine heads, teach.   Whatever you do, try to
spread the light you have.   Keep your light burning, and spread
the good news that you have to carry.   Take it for granted that
you know more about it than anybody you come in contact with,
unless it be a fellow-phrenologist.   Tell what you know, and be
assured of another thing, that you have the backing of the New
York office.   We are always glad to do for those who are trying
to help this cause along.

"I am more than pleased by the spirit of the new Alumni Association, and I hope soon to see every graduate enrolled and
that there shall arise among them the spirit of cordial union.
I am glad we have met here to-night.   I believe that this has
proved so happy an affair to all here that it will be but the first
of the annual class dinners."

The benediction, pronounced by Rev. A. C. Dill, class of '83,
closed the memorable festivities.

# FIELD NOTES.

So many of our students are making a good mark as lecturers or in the other professions that we cannot find room for extended notices of them.

We give below brief notes of some who are in the field.  We hope our friends will
report to us their plans previous to the next issue, that a note may be made for
future use.

W. G. Alexander, class of '84, is working in Utah, the home of his wife, and is finding a good field for long-continued efforts.

Eugene W. Austin, class of '78, is connected with the emigrant department in
city, where he finds abundant opportunity for observation.  He is making a very
fine collection of photographs of different types.  The collection is sure to become
of great value.

L. C. Bateman, class of '71, is doing good work contributing to the newspaper press,
in which he has acquired an enviable reputation, and also lecturing and making examinations.

Dr. J. C. Bateson attended a part of the session of the institute of '90, completing
his course, which was broken into the year before, he being obliged to leave before it
was finished.

Miss Pearle Battee, class of '90, has been working in N. J., in connection with Miss
Viola Evans, '90, where they have made good, stirring, and acceptable efforts.  She
spent the holidays with us in the city, and with the beginning of the year went to
Baltimore, where she is to spend some time.

Mr. and Mrs. Albert Bausch are working in the vicinity of New York, making a
specialty of lecturing in churches, where the proceeds are divided.

S. Brolin, class of '90, has ordered a good outfit and expects to spend the season in
Minn., where he ought to do good work among the Scandinavians.

Dr. J. L. Capen, whose portrait appears in the February number of the JOURNAL, still keeps his office open in Philadelphia.

Dr. D. Hugo Campbell, class of '74, is spending the winter in Canada, and pushing the work with his usual vigor and success. Wherever he goes he finds or awakens an interest in the work.

Rev. I. H. M. Deahm, class of '89, is continuing in his college work as a teacher and at the same time giving much attention to Phrenological matters. We wish he could take up Phrenological work to the exclusion of all other. No doubt he would find it a pleasant and profitable field.

Edgar C. Beall, class of '77, is at Cincinnati and has taken up the study of medicine as an adjunct to his Phrenological work.

Clinton E. Brush, class of '87, maintains his enthusiasm for the subject by continual application of its principles in his business.

E. E. Candee suffers from a want of health, coming from an inherited weak constitution. He hopes to be able to re-enter the field before the season is over.

Mrs. Ida P. Davis, class of '88, is in the field in Minn., and reports good public interest in her lecturing and professional work. She is very earnest and thus creates enthusiasm among the people.

S. P. Devore, class of '87, who was married during the past year, is now working westward; was at last accounts in Wyoming Ter.

Rev. David Dodds, M D., continues his pastoral work, besides conducting a department of mental science in one of the Western colleges.

Andrew A. Drury, class of '82, sends frequent orders from New England, where he finds a congenial field in which to work.

G. W. Dutton, class of '87, is editing a newspaper in Sioux City, Iowa, in which he he has a Phrenological department.

W. J. Duval, class of '89, sends occasional subscriptions for the JOURNAL and orders books from Ark., but his legal work occupies most of his time.

V. P. English, class of '86, is making an extended tour across the country and on the Pacific slope. He is one of the most earnest men in the field.

Miss Viola Evans, class of '90, spent the first part of the season with Miss Battee in New Jersey, where they met with much success. She is now in Ohio, and we hope for good reports of her work.

Ira L. Guilford, class of '81, has been out of the field for a season, but writes from Maryland that he expects to enter the field again in '91, and with renewed health will certainly do good work.

N. W. Fitzgerald, class of '85, is fully occupied with his pension and patent office in Washington, where he does a very large business; but he looks forward to the time when he can give both time and attention to Phrenology, in which he is perhaps more than ever interested.

Samuel Grob, class of '82, is interested in promoting Phrenological interests in every way he can, and at the last session of the Teachers' Institute in his county, Montgomery, Pa., he arranged for a series of lectures by Dr. Drayton which awakened a new and lively interest in the subject among the teachers.

J. B. Harris, class of '88, whose home is in Texas, made a tour through the central States very successfully, sending large orders for books. He is now back in Texas, where he looks forward to doing a good season's work.

Miss Lizzie Henderson, class of '88, is creating an interest in the subject among the Canadians, where her home is, and at the same time doing good for herself.

E. T. Hildebrand, class of '89, is at present in Ohio, where he is doing Phrenological work in teaching classes and making examinations.

Paul Howard, class of '85, is still working in the vicinity of New York, especially in the schools and colleges.

Levi Hummel, '77, is in Pa., earnestly working to interest the public in the subject, and he is doing good.

Rev. Dr. Hunter, class of '87, has just accepted a promotion to the best Methodist church in his country, at Montreal, Can. He writes us that he had hoped to retire from the ministry and devote himself to scientific work, but has yielded to the importunities of his friends to labor in that field for a short time longer; but wherever he is his interest in that subject will be felt.

David M. King, class of '67, is still living in Ohio, and during the season is actively interested in the work of the Cleveland Phrenological Institute.

W. D. Lamb, class of '89, has ordered a large outfit and is preparing to enter the field with an enthusiasm and earnestness that will insure success.

R. G. Maxwell, class of '87, is in the field in North Carolina.

Dr. Duncan McDonald, class of '67, is still on the Pacific coast, where he has large real estate interests, and doing Phrenological work.

Geo. MacDonald, class of '90, has been working with great acceptance in New England, and proposes now to remain at his home in Albany, N. Y., and divide his attention between professional work and aiding in the care and management of a good business which he had established there.

Frank Mannion, class of '79, attended the last session of the institute for another course of instruction, and proposes now to make the subject his life-work.

Miss M. L. Moran, class of '85, holds an important position in Col. Fitzgerald's patent and pension office in Washington. She applies her knowledge of Phrenology to her business in a most successful way.

Mr. and Mrs. Geo. Morris spent the last year in St. Paul and other places in Minnesota, and at last accounts were contemplating opening in Chicago, but were a little undecided between this course and the Western route. Their work of lecturing, teaching classes, and examining always bears fruit for themselves as well as for their patrons and the cause.

Howell B. Parker, class of '75, continues his academic school work in Fayette, Ga., and is ever ready to declare the important service that Phrenology has rendered to him in securing his success.

B. F. Pratt, M.D., class of '75, is still occupying the field in the West, doing considerable work in Neb.

G. P. Russell, class of '88, is in the field in Minnesota, meeting with success.

De L. Sackett, class of '89, has recently ordered supplies, and hopes to so arrange his business matters as to be able to give his attention to this subject.

V. G. Spencer, class of '90, is teaching school and making examinations, meeting with marked success in both.

D. D. Stroup, '88, was in Pennsylvania in January, 1891, doing a brisk business.

Henry E. Swain, class of '70, is still in New England working in the large cities, where he sells a great many copies of "Heads and Faces" and other books.

Dr. U. E. Traer, of Iowa, sends good orders for charts, etc., and states that he is in the lecture field for the season.

J. H. Thomas, class of '89, is making his home in Ohio, and sends occasional orders for Phrenological publications.

S. R. Vincent, class of '89, is doing good work in Oregon, and sends frequent orders for advertising matter, charts, books, etc.

Mr. and Mrs. Windsor, class of '88, were at last accounts in Texas, and, as is usual with them, having large audiences and meeting with great success.

There are many who are doing more or less in the field, but we do not hear from them sufficiently often or with enough definiteness to make any special announcement of their work.

We will try to make it of service to those who will report often in regard to their work and their plans. In the case of those who are willing to respond to calls to lecture in their own vicinity, we will be very glad to let this be known if they will let us know their P. O. address. This, of course, is done without charge.

# THE CHARTER.

### An Act to incorporate "THE INSTITUTE OF PHRENOLOGY."

Passed April 20, 1866, by the Legislature of the State of New York.

"The people of the State of New York, represented in Senate and Assembly do enact as follows:

*Section* 1. Amos Dean, Esq., Horace Greeley, Samuel Osgood, D. D., A. Oakey Hall, Esq., Russell T. Trall, M. D., Henry Dexter, Samuel R. Wells, Edward P. Fowler, M.D., Nelson Sizer, Lester A. Roberts and their associates, are hereby constituted a body corporate by the name of "THE AMERICAN INSTITUTE OF PHRENOLOGY," for the purpose of promoting instruction in all departments of learning connected therewith, and for collecting and preserving Crania, Casts, Busts, and other representations of the different Races, Tribes, and Families of men.

*Section* 2. The said corporation may hold real estate and personal estate to the amount of one hundred thousand dollars, and the funds and properties thereof shall not be used for any other purposes than those declared in the first section of this Act.

*Section* 3. The said Henry Dexter, Samuel R. Wells, Edward P. Fowler, M. D., Nelson Sizer, and Lester A. Roberts are hereby appointed Trustees of said corporation, with power to fill vacancies in the Board. No less than three Trustees shall constitute a quorum for the transaction of business.

*Section* 4. It shall be lawful for the Board of Trustees to appoint Lecturers and such other Instructors as they may deem necessary and advisable, subject to removal when found expedient and necessary, by a vote of two-thirds of the members constituting said Board; but no such appointment shall be made until the applicant shall have passed a personal examination before the Board.

*Section* 5. The Society shall keep for free public exhibition at all proper times, such collections of Skulls, Busts, Casts, Paintings, and other things connected therewith, as they may obtain. They shall give, by a competent person or persons, a course of not less than six free lectures in each and every year, and shall have annually a class for instruction in Practical Phrenology, to which shall be admitted gratuitously at least one student from each Public School in the City of New York.

*Section* 6. The corporation shall possess the powers and be subject to the provisions of Chapter 18, part 1, of the Revised Statutes, so far as applicable.

*Section* 7. This Act shall take effect immediately."

## THE FACULTY OF INSTRUCTION.

Among those who have been engaged as lecturers in connection with the Institute for many years, we may mention the following:

Nelson Sizer, the chief Examiner in the office of Fowler & Wells for forty years, lectures on the Theory and Practice of Phrenology and Physiognomy, and brings unsurpassed experience as an examiner to the instruction of students in the application of Phrenology to choice of pursuits on adaptation in marriage, the study of temperament on the living subject, health, etc.

H. S. Drayton, M. D., editor of the PHRENOLOGICAL JOURNAL, treats of Mental Science and its relations to Physiology and Metaphysics, including the lines of thought and investigation which have led up to the present state of human science.

Mrs. Charlotte Fowler Wells lectures on the History and Progress of Phrenology in America. Mrs. Wells being familiar with all that has been done to make phrenology practical, popular and useful to the home and the community, her work for students is eminently useful and always becomes memorable.

Nelson B. Sizer, M. D., Anatomy, Physiology, and Diseases of Body and Brain.

John Ordronaux, M. D., L. L. D., late State Commissioner of Lunacy, lectures on Insanity and Jurisprudence.

Rev. A. Cushing Dill, Elocution and Vocal Culture in relation to Public Speaking.

In coming to New York you should purchase a through ticket if possible, and if you have a trunk or valise which you do not need on the way, get it checked, and thus save care.

Students should prepare the means for payment of tuition and their necessary expenses during their stay in New York, *before they come.* Those who can do it should bring their funds in drafts, then they are not subject to the danger of losing their money on the way. Those who bring money can have it deposited in bank while here, thus preventing the possibility of loss.

We advise students after buying their passage tickets, to have only so much money within reach as will pay their current expenses on the way here. The balance, if not in form of draft, should be sewed into a pocket in the undergarment. Nor should students inform strangers who they are, where they come from, where they are going, or their business in New York. For in all large cities there are always men on the lookout for strangers, whose business it is to employ some cunning device to rob them. And they would rob *you.*

On landing at Jersey City from the West or South, retain your baggage check—pay no attention to agents on the train—and come to our office, 775 Broadway, above Ninth street. If you come into the city in the night, go to the St. Nicholas Hotel, 713 Broadway, corner Washington Place, two blocks from our office.

## ROOMS AND BOARD.

Boarding can always be obtained near the Institute at moderate prices. From four to five dollars a week usually cover the expense. Sometimes hygienic students club together and take rooms and procure their own food to suit themselves, and thus make a saving of expense.

We take special pains to aid students to find desirable quarters and to facilitate any purchases which they may wish to make, or give them directions as to places of interest to be visited, and the proper way to make their stay in the city safe, pleasant, and instructive.

## HEALTH IN NEW YORK.

We believe New York, with its present modern improvements for cleanliness and ventilation, is as healthy a place as there is in the land, unless it be some mountain-top. And most of our students not only maintain their health, perfectly, but gain during the course sometimes ten pounds in weight.

## OUTFIT.

Some ask us in respect to outfit. Our reply is, that one can spend from fifty dollars to two hundred dollars profitably, in the way of outfit, or can start with a very little, and add to it as he has means and feels disposed. A man can start with nothing but his hands and his tongue to work with. He may start with ten dollars in the way of apparatus and material, but he would do better with fifty dollars.

Those who contemplate visiting the city for the purpose of attending the Institute, will do well to cut out and bring this article in their pocket for reference when about to reach New York, so as to avoid confusion and mistakes.

## SECOND COURSE STUDENTS.

As an evidence of the value of the Institute course, we may mention that nearly every year one or more students return to take a second course, which is afforded to them at half price, and we notice the marked difference in second-year students, especially after they have been in the field, and learned to make practical their knowledge.

## FEES REDUCED.

The terms of instruction were four years since reduced, making them fifty dollars to all, instead of a hundred dollars for men and seventy-five for women, as formerly.

48

ANNOUNCEMENT.

# THE AMERICAN INSTITUTE OF PHRENOLOGY

Opens its annual session on the **First Tuesday of September** in each year. There is only **One Session** during the year. No person admitted for less than a **Full Term.**

This is the only institution of the kind in the world where a course of thorough and practical instruction in Phrenology is given, and nowhere else can be found such facilities as are possessed by the American Institute of Phrenology, consisting of a large cabinet of skulls—human and animal—with busts, casts, portraits, anatomical preparations, skeletons. plates, models, etc., and some teachers of fifty years' experience.

NELSON SIZER, *President.*     C. FOWLER WELLS, *Vice-President.*

HENRY S. DRAYTON, A. M., M. D., *Secretary.*

By action of the Board of Trustees, the FOWLER & WELLS COMPANY has been appointed financial and business agent. All communications should be addressed,

FOWLER & WELLS CO.,775 Broadway, New York.

## THE COURSE OF INSTRUCTION.

This consists of more than one hundred lectures and lessons covering a term of **Six Weeks**—one lesson being given each morning, and two during the afternoon.

## TOPICS EMBODIED IN THE COURSE.

**General Principles.** The philosophy of the organic constitution, its relation to mind, character, and motive ; mental philosophy, or the efforts of the best thinkers in all ages to find out the laws and operations of the mind and give their speculations the form of science. Superiority of Phrenology over every other system.

**Temperament,** as indicating quality and giving tone and peculiarity to mental manifestation, also as affecting the choice of occupation; the law of harmony and heredity as connected with the marriage relations ; what constitutes a proper combination of temperaments with reference to health, long life, tendency to talent, virtue, and vice. The subject will be largelyillustrated by subjects of real life before the class. Extended drilling of the students on this important topic.

**Phrenology.** Mental development explained; the true mode of estimating character according to phreno'ogical principles; Comparative Phrenology, the development and peculiarities of the animal kingdom; the facial angle, embodying curious and interesting facts relative to the qualities and habits of the animals; instinct and reason; the Phrenology of crime; imbecility and idiocy; the elements of force, energy, industry, perseverance; the governing and aspiring groups; the division between the intellectual, spiritual, and animal regions of the brain, and how to ascertain this in the living head; the memory, how to develop and improve it; location of the organs of the brain, how to estimate their size, absolute and relative.

**Physiognomy.** The relations between the brain and the face, and between one part of the system and another as indicating character, talent, and peculiarities, voice, walk, etc.

**History of Phrenology in America and Europe,** and the struggles and sacrifices of its pioneers in disseminating its principles, especially in this country; and its enriching influence on education, literature, domestic life, government, morality, and religion.

**Ethnology.** The races and tribes of men, their peculiarities and how to judge of nativity of race; especially how to detect infallibly the skulls of the several colored races.

**Dissection** and demonstration of the human brain; microscopic illustrations of different parts of the system in health and disease.

**Anatomy and Physiology.** The brain and nervous system; the bones and muscles; how to maintain bodily vigor and the proper support of the brain; reciprocal influence of brain and body; respiration; circulation; digestion; growth and decay of the body; exercise; sunlight; sleep.

**Objections to Phrenology,** whether anatomical, physiological, practical, or religious, will be considered; how the skull enlarges to give room for the growing brain; the frontal sinus; loss or injury of the brain; thickness of the skull; fatalism materialism, moral responsibility, etc.

**Phrenology and Religion.** The moral bearings of Phrenology, and a correct physiology; their relation to religion; home training of the young as applied to education and virtue.

**Choice of Occupations.** Special attention will be given to this branch of the subject; what organizations are adapted to the different professions and pursuits, and how to put "the right man in the right place," in actual life.

**Phrenology and Marriage.** The right relation of the sexes; what mental and temperamental qualities are adapted to a happy union and a healthy offspring, and why.

**Natural Language of the Faculties.** The attitudes, motions, carriage of the head, style of speech, from the activity of the different organs, and how to read character thereby.

**Examinations** of heads explained; practical experiments;

heads examined by each of the students, who will be thoroughly trained and instructed how to make examinations privately
and publicly; especi ully training in the examination of skulls.

**Hygiene.** How to take care of the body as to dress, rest,
recreation, food. diet, right and wrong habits; what food is best
for persons of different temperaments and pursuits; what food
tends to make one fat or lean; what feeds brain or muscle;
stimulants. their nature and abuse ; what to avoid and why.

**Psychology.** Under this head, mesmerism and clairvoy-
ance will be explained, and the laws discussed on which they
are supposed to depend.

**Heredity.** The law of inheritance in general and in particular; resemblance to parents, how to determine which parent a
person resembles; what features of face, what classes of faculties
or portions of the general build are inherited from the father
or from the mother.

**Insanity,** its laws and peculiarities; the faculties in which
different persons are most likely to be insane, and how to detect
it in a person.

**Idiocy,** its causes and how to avoid them ; its peculiarities
and how to understand them; how to detect it where the head
is well-formed.

**Elocution.** How to cultivate the voice; eloquence, how to
attain the art; careful instruction in reading and speaking with
a view to success in the lecture field.

**How to Lecture.** The best methods of presenting Phrenology and Physiology to the public; how to obtain audiences
and how to hold and instruct them; general business management in connection with the lecture field.

**Review and Examination.** Questions on all points
rel iting to the subject, which may be proposed by the students,
answered; in turn, students will be examined on the topics
taught, who will give in their own words their knowledge of
the subject. No recitations or memorizing will be required.

**How to apply Phrenology** practically in reading character by the combinations of faculties, and how to assign to each
person the true field of effort in education, business, social
adaptation, and, in short, how to make life a success and its
opportunities the means of happiness.

**Finally,** it is the aim of the instructors to transfer to students,
so far as it is possible, all the knowledge of Anthropology which
a long experience in the practice of their profession has
enabled them to acquire—in a word, to qualify students to take
influential places in this man-improving field of usefulness.

**TERMS.**--The cost of tuition for the full course, including
diploma, for ladies and gentlemen, is reduced to $50. The
lowness of the terms should insure a large class. Incidental
expenses in New York, including board, need not cost more

than $35. We aid students in getting good rooms or board. It is desirable that all who intend to be students should send in their names at an early day. For further information address, FOWLER & WELLS Co., 775 Broadway, New York.

## STUDENTS' TEXT-BOOKS.

Among the works most useful to be studied by those who wish to master Phrenology, whether at home or to prepare to attend the "Institute," we recommend the following "STUDENT'S SET," and in the order named. Be sure and learn from the Bust the location of the organs, and their nature and use from the books.

**Brain and Mind ;** or, Mental Science Considered in Accordance with the Principles of Phrenology and in relation to Modern Physiology. Illustrated. By H. S. DRAYTON, A. M., M. D. and JAS. McNIEL, A. M. $1.50.

**How To Read Character.** A New Illustrated Hand-book of Phrenology and Physiognomy, for students and examiners, with upward of one hundred and seventy engravings. By S. R. Wells. $1.25.

**The Phrenological Bust,** showing the location of each of the Organs. Large size. $1.00.

**Choice of Pursuits ;** or, What to do and Why. Describing seventy-five trades and professions, and the temperaments and talents required for each. Also, how to educate on phrenological principles—each man for his proper work; together with more than one hundred portraits and biographies of successful thinkers and workers. By Nelson Sizer. $2.00.

**Heads and Faces, and How to Study Them.** A manual of Phrenology and physiognomy for the people. By Nelson Sizer and H. S. Drayton. Oct., paper. 40c.

**Forty Years in Phrenology ;** Embracing Recollections of History, Anecdotes, and Experience. Illustrated. By Nelson Sizer. $1.50.

**New Physiognomy ;** or, Signs of Character, as manifested through temperament and external forms, and especially in the "Human Face Divine." With more than one thousand illustrations. By S. R. Wells. $5.00.

**Constitution of Man ;** Considered in relation to external objects. The only authorized American edition. With twenty engravings and portrait of the author. By George Combe. $1.25.

**Popular Physiology.** A Familiar Exposition of the Structures, Functions, and Relations of the Human System and the preservation of health. By R. T. Trall, M. D. $1.00.

N. B.—If a person already has one or more of the above books, he may order, in place of it, any other work of our publication of equal price.

Either of the above will be sent by mail on receipt of price, or the complete "STUDENT'S SET," amounting to $14.90 will be sent by express for $10.00. Address,

**FOWLER & WELLS CO., 775 BROADWAY, NEW YORK.**

# Who Should Study Phrenology?

**Young Men** who have to work their own way to eminence? It will be an advantage to them to be able to understand those with whom they come in contact, will it not?

**Young Women?** Will they not find value in being able to judge correctly the worth of young men who may pay their addresses? If they must earn their own living, will anything assist them more than the ability to measure persons correctly?

**Mothers?** Do they not need help in the proper management and training of their children? Will anything help them like understanding the peculiarities of the little ones?

**Housekeepers?** Can all servants be treated alike? Is there any science, aside from Phrenology, that will tell why they canr ot? Is it not of value to know who may be dictated to and who will e the best kind of help if orders are given as suggestions?

**Clergymen?** They must be familiar with the operations of the mind, must they not? When they can demonstrate to selfish men that they may be happier in this life as well as hereafter, by cultivating their moral natures, they will have added power, will they not?

**Lawyers?** Must they not judge their clients? Must they not be able tell the nature of witnesses and their desire to tell the truth, and also to understand each man on the jury to be able to appeal to them effectively?

**Physicians?** They must consider the constitutions and idiosyncrasies of their patients as well as their ailments, must they not?

**Teachers?** Do they find all the pupils alike? Can they tell why they are not? Will they not be aided by knowing whom to encourage and how manage the obstreperous, the dull and the precocious?

**Agents?** Will they not be aided by ability to read strangers? Will it not be an advantage to them to know with whom they may be free and social and with whom dignified and reserved, etc.?

**Managers?** They will be helped if they know before employing a person that he will prove competent, will they not? They will be aided if they can reject intelligently such as apply that are not adapted to the work in hand, will they not?

**Everybody?** If not, why not? Is there any other system of mental philosophy that will enable a man to know himself and his neighbors? Is it not of advantage to every person to possess such knowledge?

**Where?** The American Institute of Phrenology is the only place where a thorough course of instruction is given with its application to all the affairs of life. Here are the most competent instructors in the world and the largest cabinet and apparatus in existence. For full particulars address

## FOWLER & WELLS CO., 775 Broadway, New York.

# Phrenology for Business Men.

Business men who have to deal with other men, whether strangers or not, should have a full knowledge of Human Nature. Some have this intuitively, and form correct impressions, but even with these impressions, they will understand the motives of men, and know how to deal with them better, if they know why people are what they are. Others are, by nature, easily misled in their estimates of Human Character, certainly such should make this a matter of study; not only will this help to understand customers and to know how to deal with them so as to please them, but this knowledge will enable the possessor of it to handle men to his own advantage, whether in getting the best results from their work, or in knowing how to influence them to do that which is desired. Some men can be forced by a strong pressure being brought to bear, and will yield their wishes to others. Some men with large Combativeness and Destructiveness cannot be forced, but can be coaxed or influenced to do the thing desired. The man or woman who understands this subject will not make mistakes in their attempts to accomplish the desired results in life. At each session of the American Institute of Phrenology there are a number of business men who have no other motive in attending except to obtain this knowledge for application in the practical affairs of life. One of the students of a recent class, a business man residing in this city, permits us to publish his opinion as to the value of the course of instruction, as follows:—

"After attending the course of lectures of the American Institute of Phrenology for the season of 1887, I take pleasure in saying I have derived a great deal of practical benefit from them, and am very much surprised that a subject which has been so ably and elaborately explained and taught should receive so little attention from practical business men. Although I have been compelled to neglect pressing business engagements to attend this course of lectures, at the same time I feel satisfied that they have well repaid me for the time I have devoted to them, and as far as the expense is concerned, I consider it one of the best investments I ever made.

A very important thing to every business man is an education which will enable him to put the right man in the right place, and I know of no course of instruction that will compare with Phrenology in this important matter."

CLINTON E. BRUSH.
Manager Chicago Corset Co., 402 Broadway, N. Y.

The man who best understands Human Character, by study or experience, is the one who will work most successfully and with the least wear and tear of mind and body, among men, in the transaction of business. The American Institute of Phrenology was organized for the purpose of giving this instruction, and we will be glad to give any information that may be desired relating to the matter. Address

## FOWLER & WELLS CO.,

*775 Broadway, New York.*

# THE PHRENOLOGICAL MUSEUM.

## 775 BROADWAY, N. Y.

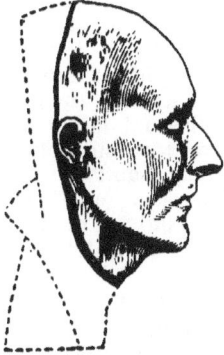

THIS is the only collection of the kind and contains CASTS from Life and BUSTS of hundreds of celebrated people in whom the public are interested. Among others the following of many of these no duplicates are in existence, and the originals have been procured at great expense.

Napoleon, I.; Webster, Lincoln, John C. Calhoun, Henry Clay, Grant, Voltaire, Thos. Paine, Guiteau, Lord Byron, Dante, Bryant, Edison, Huxley, Gen. Hancock, Gerritt Smith, Chastine Cox, Rugg, Idiot Family (Hillings), Laura Bridgeman (deaf, dumb, and blind), Cuvier, Prof. Morse

MASK OF NAPOLEON I.    Edwin Forest, Kean, Walter Scott, Robt. Burns,
John Quincy Adams, Gall, Spurzheim, Dr. Tanner, McClellan, Henry Ward Beecher, Dr. Cox, Dr. Valentine Mott, Mrs. Gottfried (murderess), Salmon P. Chase, Houdans, Washington, Sumner, Cardinal McClosky, Thos. H. Benton, Black Hawk, Aaron Burr, Alex. Hamilton, O'Connell, Gilmore, Washington Irving, Mendelssohn, Beethoven, Oliver Cromwell, Wordsworth, Colridge, Tom Moore, Sir Isaac Newton, Patty Cannon, a case water-brain; Hara Waukay (New Zealand Cannibal), Elihu Burritt, Gen. Scott, Franklin, Greeley, Seward, Jas. Fisk, John Kelly, Rich. B. Sheridan, Lord Chatham, Robt. Bruce, Wm. Pitt, Rev. John Pierpont, Robt. Dale Owen, Dr. Hahneman, Dr. Casnochen, Dr. Newman, Dr. Alcott, Chas. Dickens, Raphael, Pres. Barnard, Dio Lewis, Pere Hyacinth, Flat-headed Indian, Captain Cook, Osceola, Horace Mann, Dr. Saml. Howe, Dr. Trall, Francis Wright, Frederick, the Great, Dean Swift, Martin Van Buren, Milliard Fillmore, Zack Taylor, Jas. K. Polk, Lafayette, Rev. Leonard Bacon, Neanderthal, Man and Gorilla, Mad Malibran, Sylvester Graham, John C. Fremont, Dupuytren, Dr. Chalmers, and many others.

Among the many portraits and sketches in Oil and Crayon are the following: Gladstone, Parnell, Dr. McCosh, Wm. M. Evarts, Wendell Phillips, C. Vanderbilt, W. Vanderbilt, Geo. Peabody, Fred Douglass, John Brown, Henry Wilson, Mark Lemon, H. Dana, Stewart Mills, Rich. Cobden, Tupper, Rosa Bonheur, Robt. E. Lee, Gustave Dore, McMahon, Emerson, Peter Cooper, Stanton, John B. Gough, Maximilian, Alex. H. Stephens, Jefferson Davis, Christine Neilson, Michael Angelo, Schuyler Colfax, Brigham Young, Francis Wayland, Dr. Morgan Dix, Disraeli, Bancroft, Goethe, Cæsar, Wm. Lloyd Garrison, S. H. Tyng, The Duke of Wellington, Edgar A. Poe, Thier, Agassiz, John Jacob Astor, Pope, Alexander VI., Dr. Guthrie, Lucretia Mott, Stephen Girard, Mrs. Garfield, De Lesseps, Gov. Francis Train, Grace Greenwood, Talmage and many others.

These with many others are catalogued and on free exhibition. You are cordially invited to spend an hour, more or less, at our rooms at any time, where a competent person will freely answer your inquiries.

FOWLER & WELLS CO., Phrenologists and Publishers,
775 BROADWAY, N. Y.

# THE STUDY OF HUMAN NATURE.

In the study of Human Nature we find one person who is full of fire and needs guidance and restraint, another is timid and diffident, lacking in force and fortitude, and needs encouragement; another is too sentimental and should be taught the need of a more practical life; another is given to sordid greed, and worships, if not the "Golden Calf," the gold that might make one; another requires advice as to diet and daily habit and hygiene; another is precocious, too imaginative, too intellectual, and needs ballasting and instruction in the way of daily habit and economic duty; another is imperious, irascible, and impatient; another is inclined to dissipation; another desires to know what he can do best, what kind of trade, business, or profession his talents, constitution, and aptitudes best fit him for; another is broken down by over-work or over-study and needs information as to the cause and cure of the trouble.

Occasionally there may be a man so harmonized in body and mind, so smoothly related to life that he does not need help from physician, phrenologist, or life insurance company. Most people, however, need something to fill out their deficiencies or restrain their excesses, or to guide their forces. As a locomotive carries its headlight in its front and illuminates a mile or two of track in advance of itself, so a proper description, phrenologically and physiologically, is calculated to illumine the pathway of life, and if it does not make the grade easier it makes the transit more safe and sure.

The purpose of a Phrenological Examination is to study the Temperament, or constitution in relation to health, talent, and character, and how the different vital organs are developed and act with each other in the promotion of physical and mental harmony and power. Next the size of the brain and the quality which the temperament gives it; then the developments of the different groups of organs; those of intellect, perception, memory, reason; those of force and energy, of policy, prudence, thrift, ingenuity, taste, refinement; those of aspiration, pride, self-reliance, ambition; those of social power and affection; and last, the strength and tendency of the moral sentiments.

We do not, as some suppose, look for little hills and hollows or bumps, but at the distance or length of fiber from the basilar center of the brain.

We also describe the adaptations of each person for given pursuits, in which their abilities can be used to the best advantage. We teach parents how to understand their precocious children who need prudent care to save them, also how to train turbulent and vicious children, to bring their moral and intellectual powers into the ascendant.

Our cabinet, containing hundreds of busts, casts, portraits and sketches of men and women, noted and notorious, from all classes, including statesmen, soldiers, lawyers, divines, inventors, philanthropists, etc., with murderers, pirates, and others from the lower walks of life, with many recent additions, is catalogued and free for the inspection of visitors daily. Citizens and strangers will find this one of the most pleasant places in the city in which to spend an hour.

Our rooms, centrally located at 775 Broadway, are near various lines of horse cars and stations on the elevated roads. To persons at a distance, and those who do not find it convenient to visit our office, we would say very satisfactory examinations can be made from properly taken pictures and measurements which can be given. For full particulars in regard to this, send for circular called Mirror of the Mind. Address
FOWLER & WELLS CO., Publishers. 775 Broadway, New York.

# CHARACTER FROM PHOTOGRAPHS.

MANY persons who reside at so great a distance that they can not visit us, desiring to avail themselves of our professional services, have written to us enclosing photographs, requesting our opinion of the character, talents, and proper pursuits of the originals.

These requests becoming very numerous, and the likenesses generally being taken in a manner not adapted to the purpose, we deemed it necessary to prepare a circular giving full instructions how likenesses should be taken for examination ; also rules for the measurement of head and body, and such other points of information as would form a basis of judgment in regard to temperament, constitution, and health.

This circular, called "Mirror of the Mind," is illustrated by engravings showing the forms of many heads, with full directions for those desiring descriptions of character.

Thousands have availed themselves of this method of learning their true character, and to what profession, trade, or occupation they are adapted ; and not a few have been saved from bad habits and wrong pursuits, as well as from unfavorable social and domestic alliances, by sending the portraits of persons of whose real characters they desired to know more than they had the time and opportunity to learn in the ordinary way.

Parents consult us in regard to the choice of pursuits for sons, whether educational, mechanical, or agricultural ; or for daughters who must make their own way in the world, and who would know whether in a trade, art, or teaching they would be most successful.

Many people are broken down in health and constitution, and need plain advice as to the proper means of recovery. Their physicians do not always tell them how to escape from their morbid conditions, because not employed to explain the case, but to treat and cure the patient. We aim to instruct the applicant, when necessary, as to the right mode of living to get rid of morbid conditions, and how to retain health and vigor by normal means.

We have received likenesses for examination from English settlers at the Cape of Good Hope in South Africa, from New Zealand and Australia, from the West Indies, from England, Scotland, Canada, Mexico, and scores of them from Oregon, California, and the Rocky Mountain settlements, as well as many from persons at shorter distances, yet so far that the cost of coming to New York would be far more than the cost of our professional services.

The circular, "Mirror of the Mind," alluded to before, explains terms, etc., and will be sent promptly to all who request it.

We have numerous letters testifying to the accuracy of these delineations, and the great practical benefit derived from the advice and instruction imparted.

## A CASE IN POINT.

A gentleman, who was a stranger to us, called at our office with the photographs of a gentleman and a lady, which he desired us to examine carefully, and to write out our opinion of the character of each, and more particularly that of the gentleman, and to give our opinion as to the adaptation in marriage of the parties, the lady being his daughter. The gentleman did not tell his name or residence, or that of the parties in interest. We promised to have the matter ready in a few hours, and he retired. We then proceeded to prepare the statement, in which we described the young man as selfish, tyrannical, and inclined to be immoral, and quite unsuited to the lady. When the gentleman called for the document, he took it, sealed, and left without reading it.

About a month afterward we received a letter from the father, addressed to the examiner which we copy :

"NELSON SIZER :—*Dear Sir*—In the latter part of March last, I was in the office of FOWLER & WELLS, and left with you two photographic likenesses (of a young man and young woman), to be examined in regard to their relative fitness for union in matrimony—more especially the young man. The study of the description I obtained from you, coupled with some recollections I have of his habits and ways, led me to the conclusion that your delineation is, in every way, true and to the point. Thanking you a thousand times for the favor conferred on me, which I consider more in the light of a friendly act than otherwise,

"I remain, very truly yours, —— —— "

Two years afterward the young lady called, made herself known, and warmly thanked us for having saved her from a sad misalliance.

From another we have the following, written by a fond and anxious parent :

"FOWLER & WELLS : *Chicago, Ill.*, July 20, 1877.

"I have just received the 'description of character' written by you, for my son, from photographs, and it would seem that you had known him from the cradle. He is peculiar ; a kind of mystery, but you describe him truly. Your advice as to his future business appears to be correct, for he has manifested talent in that direction. We desired to educate him for a profession for which you say he has but little ability. We shall follow your advice in his case, and he is delighted that you encourage his preference.

"Thankfully yours, S. H."

All letters of inquiry should contain stamp for postage, and be addressed to

# FOWLER & WELLS CO., Publishers, 775 Broadway, N. Y.

# CHARTS FOR PHRENOLOGISTS.

*How to Read Character.* The New Illustrated Hand-book of Phrenology and Physiognomy, for Students and Examiners, with a Chart for recording the sizes of all the different Organs of the Brain, in the Delineation of Character, with upwards of 170 engravings. Adapted to the use of Examiners, to be used as a Chart, and for learners, in connection with the "Phrenological Bust." Price, muslin, retail, $1.25; wholesale, 62 1-2 cents; paper covers, retail, $1, wholesale 50 cents.

*Self-Instructor in Phrenology and Physiology.* Illustrated with over 100 Engravings, together with a Chart for the recording of Phrenological Developments for the use of Phrenologists. Price, muslin, retail, $1.00; wholesale, 50 cents; paper cover, retail, 50 cents; wholesale, 25 cents. A new revised and enlarged edition just published.

*Wells' New Descriptive Chart,* for the use of Examiners, giving a Delineation of Character. Price, paper, 25 cents; $2.00 per dozen, by mail; $10.00 per hundred, by express.

Bound in flexible cloth, price retail, 50 cents; wholesale, 25 cents.

This is the best small and cheap Chart ever published; is arranged with a Table for marking, the same as the larger Charts, and is used more extensively than any other.

*Synopsis of Phrenology,* and Chart describing the Phrenological Developments, for the use of Examiners and Lecturers. Price 10 cents; $1 00 per dozen, by mail; $5.00 per hundred, by express.

The above constitute our blank Charts for the use of Practical Phrenologists, and afford a variety in size and cost that will suit all cases and circumstances. The books are well made, and, as will be seen, we make liberal discount to the profession.

Lecturers can make it pay to sell our publications, and should in all cases do this. We give to them our best terms to agents, and it is believed all would meet with some degree of success, if a stock of books were carried. Subscriptions may also be taken for the PHRENOLOGICAL JOURNAL. Names can be sent on at once, retaining the commission allowed, 50 cents, and the JOURNAL will be sent.

All orders must be accompanied by the amount in draft on New York, Express Order, Post-Office Order, Postal Note or in Registered Letter, and addressed to

## FOWLER & WELLS CO., Publishers, 775 Broadway, New York.

# WORKS BY NELSON SIZER.

**Choice of Pursuits**; or, **What to Do and Why**, describing Seventy-five Trades and Professions, and the Temperaments and Talents required for each; with Portraits and Biographies of many successful Thinkers and Workers. By NELSON SIZER, Associate Editor of the "PHRENOLOGICAL JOURNAL," President of, and Teacher in, the "American Institute of Phrenology," 12mo, extra cloth, 508 pp. $2.00.

This work fills a place attempted by no other. Whoever has to earn a living by labor of head or hand, can not afford to do without it.

**How to Teach According to Temperament and Mental Development**; or, Phrenology in the School-room and the Family. With many illustrations. 12mo, extra cloth, 351 pp. Price, $1.50.

One of the greatest difficulties in the training of children arises from not understanding their temperament and disposition. This work points out the constitutional differences, and how to make the most of each.

**Forty Years in Phrenology.** Embracing Recollections of History, Anecdote, and Experience. 12mo, extra cloth, 413 pp. Price, $1.50.

The volume is filled with history, anecdotes, and incidents pathetic, witty, droll, and startling. Every page sparkles with reality, and is packed with facts too good to be lost.

**Heads and Faces; How to Study them.** A new Manual of Character Reading for the People, by Professor Nelson Sizer and Dr. H. S. Drayton. It tells all about the subject and contains 200 pages, 250 striking illustrations from life. Paper, 40 cents; cloth, $1.

**Thoughts on Domestic Life**; or, **Marriage Vindicated and Free Love Exposed.** 12mo, paper, 25 cents.

**The Education of the Feelings and Affections.** By Charles Bray. Edited, with Notes and illustrations from the third London edition, by Nelson Sizer. 12mo, extra cloth, $1.50.

**Tobacco; Its Effects on the Human System**, Physical, Intellectual, and Moral. By Dr. William A. Alcott. With Notes, Additions, and Illustrations by Nelson Sizer. 151 pp. Paper, 25 cents.

**Tea and Coffee; Their Effects on the Human System**, Physical, Intellectual, and Moral. By Dr. William A. Alcott. With Notes, Additions, and Illustrations, by Nelson Sizer. Price, 25 cents.

**Self-Reliance or Self-Esteem as an element in Human Character,** its uses and culture, 10 cents.

**On Choice of Occupation;** or, my Right Place in Life, and How to Find it, 10 cents.

Sent by mail, postpaid, to any address. Agents wanted. Address

FOWLER & WELLS CO., Publishers, 775 Broadway, New York.

DR. FRANCIS JOSEPH GALL.

## THE FOUNDER OF PHRENOLOGY.

We have just published a new life size lithographic portrait of the great founder of phrenology. It is made from a very rare copperplate engraving and represents him as he appears above, with all the added strength that is given in making it life size. On heavy paper, 20 x 23 inches, a rich tinted print. Price, $1.00 by mail, postpaid. Address

# FOWLER & WELLS COMPANY,

*775 Broadway,*                                        *NEW YORK.*

Every phrenologist should have it, and will be pleased with it.

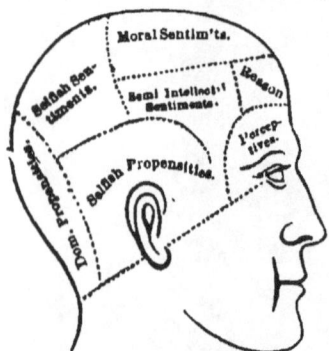

# A Natural System of Elocution and Oratory.

Founded on an analysis of the Human Constitution, considered in its three-fold nature, Mental, Physological, and Expressional. By THOMAS HYDE and WM. HYDE. Illustrated. 12mo. Extra Cloth, $ 2.00. Library, sprinkled edges, $2.50. Half Turkey Morocco, polished marbled edges, gilt back, $3.25.

We call the attention of Clergymen, Lawyers, Teachers and Scholars to this new book on Elocution and Oratory, as one of the most scholarly, original and inspiring books ever written on the subject of Eloquence. It begins a new era in oratorical instruction, since it supplies what has long been needed, and the want of which has retarded up to the present moment the progress of oratorical instruction, namely, a scientific exposition of the emotional and intellectual elements, which are at the basis of persuasive oratory. It covers the entire field of the art of eloquence, and is so full and clear in its statements and definitions that students can master this noble art by its aid without a teacher. In truth, the book is so full of instruction, thought and experience, that one can learn from reading it what it would cost him several hundred dollars to obtain through the channels of professional instruction. And, besides, as the book is original in its system, matter, and plan, what it contains can not be obtained from any one professor of the art. It is a book which ought to be in every one's possession and be studied slowly and carefully, for every perusal of it will reveal something new, and strengthen and improve the talents of speech. The authors have brought to bear ripe experience and zealous study in its production, and hence it is unsurpassed as an exhaustive, practical and comprehensive treatise. There are no superficial glimpses given of oratorical truths ; no mere verbal enunciations of graceful positions of body, or tricks of voice intonations, but the authors go to the root of the matter and unfold the germinal thought and oratorical passion ; show how such thought expresses itself in look, voice and gesture, and how each may be cultivated. It teaches the clergyman how to improve the composition and delivery of his sermons ; the lawyer, the most persuasive arrangement of the arguments and details of his plea, and the platform orator, how to make his discourse suit the needs of a popular assembly, and the teacher of elocution the method of instruction best adapted to awaken and stimulate the natural endowments of his pupils. The book is invaluable to teachers in Public Schools, since it gives a careful, clear, and exhaustive analysis of every vocal element with its correct pronunciation. Such analysis, with the many other valuable suggestions it contains, will enable the teacher to drill his pupils successfully in articulation and pronunciation. The book, besides teaching in a thorough manner all that is essential to oratory, unfolds in a practical way all that is embraced under the term elocution. Public readers and actors can learn from this book more about the natural training and developing of voice and character impersonation than from any book now before the public.

Sent by mail on receipt of price. Agents wanted.

Address, **Fowler & Wells Co., Publishers, 775 Broadway, New York.**

# Right Selection
## in Wedlock.

### MARRIAGE NOT A FAILURE.

#### BY NELSON SIZER.

Marriage is a failure only when the persons are not properly mated. and this is likely to be the case only from a want of knowledge. In this work Prof. Sizer tells who should marry and who should not, giving portraits to illustrate the Temperaments and the whole subject fully. The right age to marry, the marriage of cousins, and many other questions of interest are considered. Price, only ten cents, by mail, postpaid.

*ARE THEY WELL MATED ?*

# Resemblance
## to Parents
### AND
## HOW TO JUDGE IT.

#### BY NELSON SIZER.

This work gives practical instructions for judging inherited resemblances. By its aid students may learn to tell at a glance which parent a person resembles and correctly infer much concerning the character. The work is illustrated by forty-seven cuts, and is sent by mail, postpaid, on receipt of price, only ten cents.

*LIKE FATHER OR MOTHER ?*

# Getting Married and Keeping Married.
## BY ONE WHO HAS DONE BOTH.

Under the titles "Finding a Mate" and "Keeping a Mate" the author gives points of interest to both married and unmarried. Those who wish to be loved and those who wish some one to love, will find numerous suggestions of value in its pages and illustrations. By mail, postpaid, only ten cents.

*FOWLER & WELLS CO., Publishers, 775 Broadway, N.Y.*

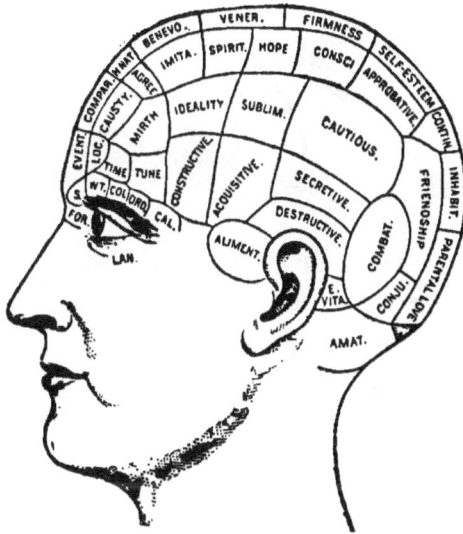

# Men and Women Differ in Character.

[PORTRAITS FROM LIFE IN "HEADS AND FACES."]

No. 1. James Parton.
No. 2. A. M Rice.
No. 3. Wm. M. Evarts.
No. 4. General Wisewell.
No. 5. Emperor Paul of Russia.
No. 6. George Eliot.
No. 7. King Frederick the Strong.
No. 8. Prof. George Bush.
No. 9. General Napier.
No. 10. Otho the Great.
No. 11. African.

## IF YOU WANT SOMETHING

that will interest you more than anything you have ever read and enable you to understand all the differences in people at a glance, by the "SIGNS OF CHARACTER," send for a copy of

## HEADS AND FACES; How to Study Them.

A new Manual of Character Reading for the people, by Prof. Nelson Sizer, the Examiner in the phrenological office of Fowler & Wells Co., New York, and H. S. Drayton, M.D., Editor of the PHRENOLOGICAL JOURNAL. The authors know what they are writing about, Prof. Sizer having devoted nearly fifty years almost exclusively to the reading of character and he here lays down the rules employed by him in his professional work. It will show you how to read people as you would a book, and to see if they are inclined to be good, upright, honest, true, kind, charitable, loving, joyous, happy and trustworthy people, such as you would like to know.

A knowledge of Human Nature would save many disappointments in social and business life.

This is the most comprehensive and popular work ever published for the price, 25,000 copies having been sold the first year. Contains 200 large octavo pages and 250 portraits. Send for it and study the people you see and your own character. If you are not satisfied after examining the book, you may return it, in good condition, and money will be returned to you.

We will send it carefully by mail, postpaid, on receipt of price, 40 cents, in paper, or $1 in cloth binding. Agents wanted. Address

## FOWLER & WELLS CO., Publishers, 775 Broadway, New York.

www.ingramcontent.com/pod-product-compliance
Lightning Source LLC
Chambersburg PA
CBHW022005190326
41519CB00010B/1392